the VEGETABLE GARDENER'S GUIDE *to* PERMACULTURE

the VEGETABLE GARDENER'S GUIDE to PERMACULTURE

Creating an edible ecosystem

Christopher Shein with Julie Thompson

Timber Press
Portland • London

Page 2: Edible permaculture gardens are abundant polycultures, mixed plantings that occupy all available space, both vertical and horizontal.

Photo credits appear on page 271.
Line drawings by Sonya Binnewies

The information in this book is true and complete to the best of the author's knowledge. All recommendations are made without guarantee on the part of the author or publisher. The author and publisher disclaim any liability in connection with the use of this information.

Published in 2013 by Timber Press, Inc.
The Haseltine Building
133 S.W. Second Avenue, Suite 450
Portland, Oregon 97204-3527
timberpress.com

6a Lonsdale Road
London NW6 6RD
timberpress.co.uk

Printed in China
Book design by Laken Wright
Second printing 2013

Library of Congress Cataloging-in-Publication Data
Shein, Christopher.
The vegetable gardener's guide to permaculture : creating an edible ecosystem / Christopher Shein
with Julie Thompson. -- 1st ed.
p. cm.
Includes bibliographical references and index.
ISBN 978-1-60469-270-9
1. Permaculture. 2. Vegetable gardening. 3. Food crops. I. Thompson, Julie, 1962- II. Title.
S494.5.P47S54 2013
631.5'8--dc23

2012013889

A catalog record for this book is also available from the British Library.

CONTENTS

PREFACE

I first became acquainted with the Australian concept of permaculture when I was in college. Ever since then I have tried to apply its ecological gardening techniques in my various roles as a community gardener, a CSA (community supported agriculture) farmer, and a compostmeister at Linnaea Farm in British Columbia. Now I run my own permaculture landscaping business and have taught permaculture for more than ten years at Merritt College in Oakland. I've created more than a hundred gardens, all of them inspired by permaculture principles and ethics.

Permaculture is an ancient yet cutting-edge technology. The ethics, principles, techniques, and strategies it employs are inspired by indigenous land practices around the world. My travels in Mexico and Central America and my work with the permaculture community have convinced me that we need to try to reweave the web of life into whole cloth. Permaculture not only aims to make the soil productive, but also to make enough room for everyone to come to the table and eat. My own five-year-old permaculture garden feeds me and my family every day, as well as many other friends and neighbors.

This book is a practical guide to basic ecological literacy and permaculture gardening. I have tried to break down the techniques and language of permaculture to show that any gardener can be a positive asset to the interconnected web of life. Planting a permaculture garden is a dream for many people who have even a small amount of land, and I believe that permaculture is a viable ecological design strategy suited to anyone's backyard—or even to a front yard, rooftop, balcony, neighbor's garden, school garden, or community garden.

Making ecological gardens is about working less hard, but smarter. My favorite quote about permaculture, from Bill Mollison, is that it is "thoughtful and protracted observation, not thoughtless and protracted labor." I hope this book helps you to grow your own permaculture garden, and that the experience is satisfying. Happy (permaculture) gardening.

CHRISTOPHER SHEIN
BERKELEY, CALIFORNIA

Bountiful plantings, mulched pathways, and insect-attracting perennials and shrubs create a welcoming entrance to this permaculture garden.

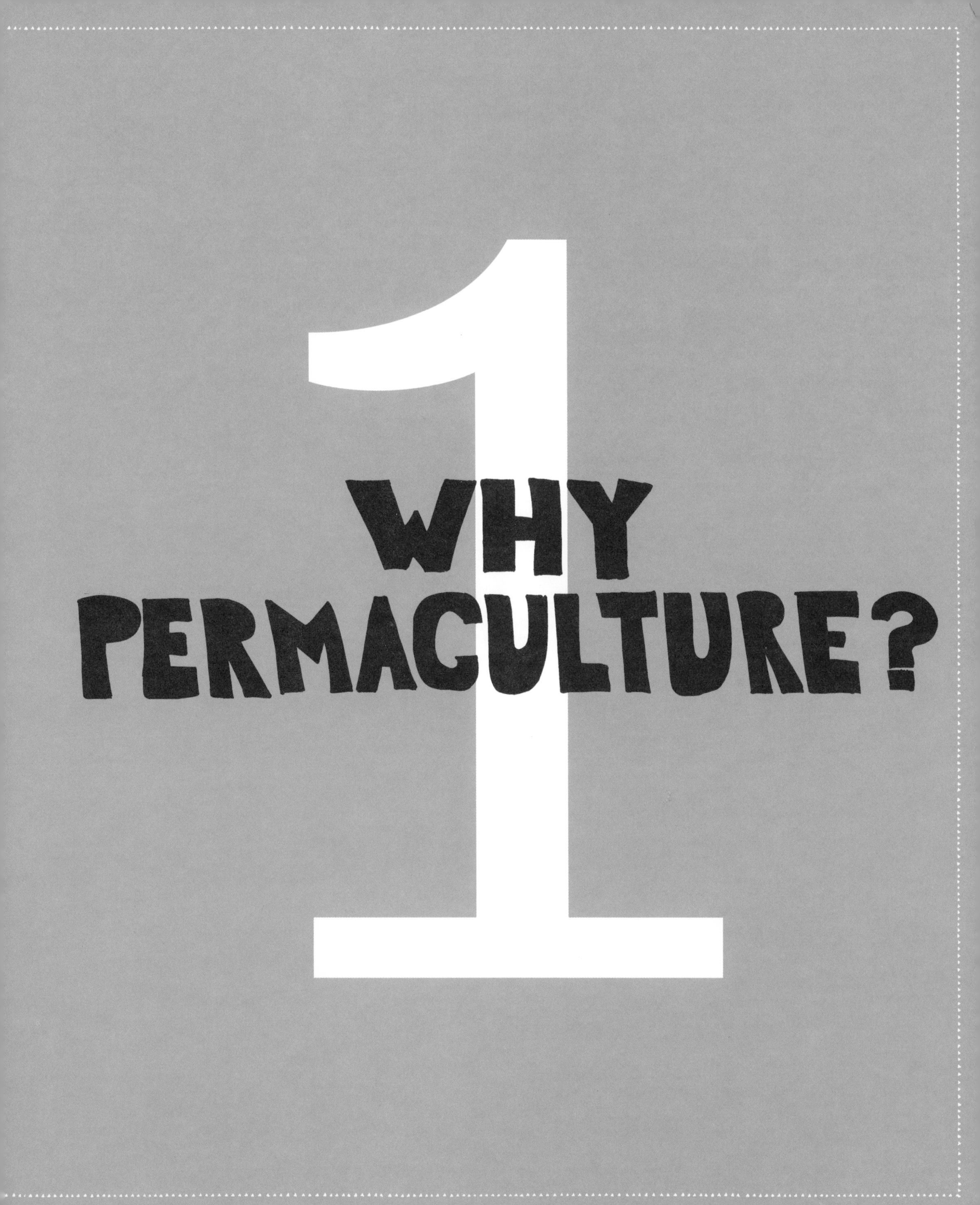
1
WHY
PERMACULTURE?

What Is Permaculture?

WHAT VEGETABLE GARDENER wouldn't like to grow more food in less time and for less money? That's exactly what permaculture offers. Instead of relying on backbreaking work, fertilizers, herbicides, and pesticides, the permaculture gardener uses methods that build healthy soil, smother weeds, and promote plant life, while recycling waste products from the garden. Whether you are a beginner gardener just starting to grow your own food or an experienced grower who wants to increase your yields, you'll find that permaculture offers design tools and growing techniques that will help you create an abundance of food for your family and friends while saving you effort in the garden.

Formalized in the late 1970s in Australia by Bill Mollison and David Holmgren, permaculture combines sustainable agriculture, landscape design, and ecology (the name is drawn from the terms permanent agriculture and permanent culture). It is an approach that encourages the home gardener to work with nature rather than against it to design a garden that thrives with minimal intervention. Although terms like *hugelkultur*, fruit tree guilds, and food forest may seem unfamiliar now, these are in fact simple concepts that can be implemented in any edible garden.

Permaculture has much in common with organic gardening, but it is a different approach. Natural ecosystems are the model, so plants are placed in mutually beneficial plant communities. There is an emphasis on perennial plants over annual ones, and permaculture gardeners grow many crops at the same time in the same location. There

Permaculture techniques can be scaled up or down to fit any size garden, from an urban balcony to a rural homestead. Fruit tree guilds, hugelkultur *beds, and polyculture plantings are all essential components. Chickens help to maintain the food forest.*

Do-Nothing Farming

Masanobu Fukuoka was a major philosophical influence on the founders of permaculture. An agricultural scientist who gave up his life as a researcher to return to his family farm, Fukuoka developed a natural farming technique that he called do-nothing farming. His first question was, "What can you not do?" This kind of farming doesn't depend on the plow to turn over the soil, or call for the spraying of chemical fertilizers and pesticides. Instead, it relies on biological systems such as beneficial insects, cover cropping, and keeping small livestock like chickens and ducks.

In Fukuoka's model, cover crops like perennial white Dutch clover are rotated with grain crops like barley, rice, and wheat. Seeds are formed into balls with clay and compost, then broadcast into the fields. After the grains are harvested, the straw is returned as mulch for the next crop. Fukuoka's book, *One Straw Revolution*, outlines his techniques and philosophy. My favorite Fukuoka quote is, "If we throw Mother Nature out the window, she comes back in the door with a pitchfork."

Clover was widely planted by Masanobu Fukuoka to attract beneficial insects and improve the soil.

Gardening wisdom can be passed from one generation to the next. My mother was my first gardening teacher. Here she is planting beet seedlings while holding her newborn granddaughter in a sling.

are ongoing recycling and re-use projects throughout the garden, such as water harvesting. And permaculture does not advocate plowing and digging the soil, but rather building it up over time with no-till methods.

Permaculture and Food

My mother was my first gardening mentor. I remember when I was growing up in Ann Arbor, Michigan, she tended vegetable patches beside the porch as well as in different community gardens. Those early years spent around vegetables gave me a great appreciation for their immense variety, both in the garden and on the table. To me, there's nothing better than a meal cooked with fresh vegetables picked directly from the garden minutes before eating. Not only does fresh produce taste better, but it is also more nutritious.

Permaculture is a perfect match for edible gardeners because in addition to creating a more sustainable and responsible garden and community, it also leads to lots of great-tasting food. Successful edible gardening relies on well-prepared soil, ample moisture, minimum weed competition, the right choice of plants, and proper timing. The amazing thing about permaculture is that it allows you to meet these needs with a minimal investment of time and money. That's because the best long-term solution to growing abundant food is to garden in ways that enrich the garden's resources rather than depleting them.

The first step in permaculture is intelligent garden design. Typical residential landscape designs with large lawns and veggie plots relegated to a far corner require a lot of maintenance and are not very efficient. Permaculture uses techniques that have been adapted from indigenous peoples around the world—such as layering and stacking—to help maximize every available growing surface: backyards, front yards, curb strips, decks, balconies, fire escapes, rooftops, along walls and fences, in neighbors' yards, and at community and school gardens. As I have seen from my own garden, these tools help create a maximum edible yield in whatever space you have available. If you are living in cold climates, permaculture methods can be applied to starting seeds indoors, making use of cold frames, hoop houses, and greenhouses to add even more growing space. In warm climates, you can use permaculture to create shade and to harvest water.

Next, permaculture finds ways to repair even the poorest soils, even if previous generations left the soil in an unhealthy condition. Dirt is the basis of all good growing. Without healthy, biodiverse soil, you cannot grow healthy and resilient plants. Rather than tilling and digging as in conventional farming and gardening, permaculture gardeners use techniques that add fertility and encourage biological activity in the soil in ways that mimic the natural soil food web.

Finally, permaculture is based on the common-sense idea of eating what grows well locally and celebrating what is in season. Baby boomers will remember when getting an orange in your Christmas stocking was a big deal. That's because oranges were

Creating vertical surfaces to expand your planting choices is a way to grow as many different plants as possible, as shown by this innovative trellis in Brooklyn, New York.

ABOVE
Permaculture gardens emphasize crops that are well adapted to the local conditions. These cabbages will taste sweeter after a frost and can be stored in the garden protected by mulch and snow cover.

OPPOSITE
Joseph Davis of City Slicker Farms in Oakland, California, gets some help from a member of the community to plant seedlings. Sharing resources and work is integral to permaculture.

grown only in places like Florida and California, and were exotic and expensive. Now that so much of our food is routinely shipped thousands of miles, we've lost a connection we used to have with local farmers and food. Permaculture encourages us to celebrate local producers by eating local, seasonal produce, and by preserving and sharing the bounty. In our own gardens, we feature a variety of plant choices based on what is best adapted to the particular garden and the tastes and needs of those who tend it.

In addition to the practical aspects of this system, it's important to realize that permaculture is more than just a way to grow plants. It's an ethical approach to growing food that reconnects us to our farming traditions. Although it's a newer system, it's based on cultural traditions that have been supplanted by industrial agriculture and fast food. It can be said that permaculture is a ten-thousand-year-old, cutting-edge technology that teaches us to grow crops in a sustainable way. The beauty of permaculture is that it embraces both traditional pre-industrial agriculture and influences from other cultures. It returns us to the model of small-scale growing, when resources were shared in the community, and the garden itself is part of the larger ecosystem.

HARD
WEAR

TREE MODEL

Looking through the lens of the tree model helps us to understand how permaculture's guiding principles relate to the design process. The whole is rooted in thoughtful and protracted observation; permaculture ethics are the driving force of our assessment, vision, concept, and actions. The tree's roots also represent the soil food web (the community of animals, plants, and fungi), while leaves are solar collectors and sources of food; they filter the wind, dust, pollen, and noise; and they eventually become mulch and compost.

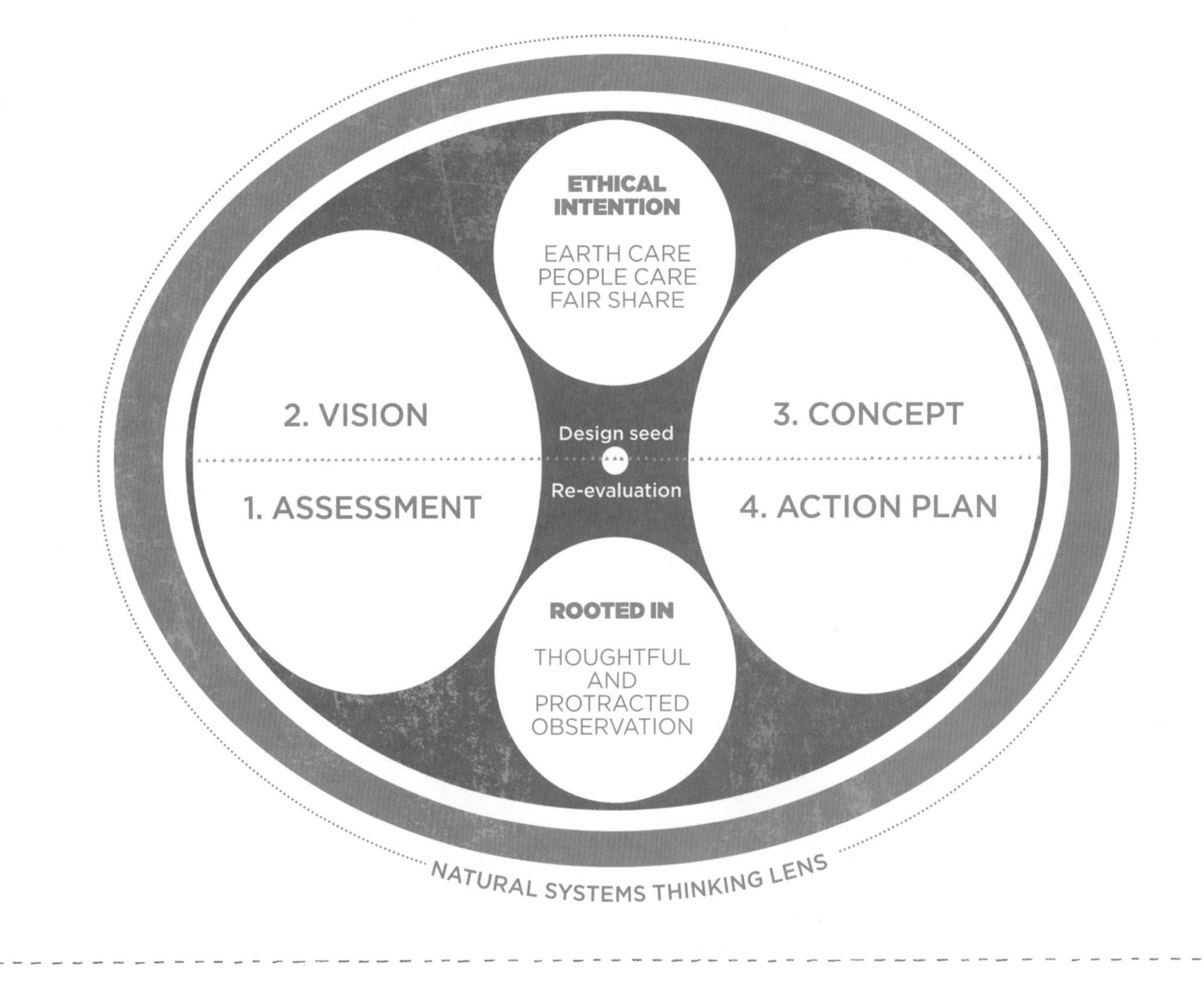

THE TREE MODEL

I like to use a tree as a theoretical model for the permaculture design process, because it helps me to view the bigger picture. I am near-sighted, and when I put on my glasses I can see distant objects in greater detail. Similarly, the lens of permaculture makes it easier to recognize patterns and beneficial connections that are not immediately apparent. The tree model refers to the fundamental guiding principles of permaculture: interrelated and connected systems. Trees start from seeds, grow, mature, produce more seeds, die, and proliferate into more trees.

The design seed, or idea, starts at the middle point of the tree model and germinates into the full design. Your idea is the inspiration to grow as much of your own food as possible. Using permaculture techniques, you will watch your garden sprout, grow, bloom, fruit, and then go to seed. You will see that gardens evolve over time, as you plant your favorites and weed out the others. Although there will be failed experiments, the permaculture gardener sees these as normal events in the dynamic process of all living environments. Mistakes are a learning tool and a sign you're continuing to improve your process.

This oak tree starts as an acorn and grows into a seedling and then a mature tree, collecting sunlight and filtering pollution. Ultimately the tree is recycled by fungi and returned into soil for future trees. The permaculture tree model is based on these natural patterns: design seed, assessment, vision, concept, action plan, and re-evaluation.

The Ethics of Permaculture

Permaculture is built on a foundation of three ethical principles: caring for the planet (earth care), caring for others (people care), and sharing abundance (fair share). These community-based principles reflect the values of many traditional cultures that look out for the interests of everyone in the group, as well as the interests of the overall community and of the planet itself.

EARTH CARE

Earth care asks us to recognize the living soil food web as essential to all life, including our own. In the words of David Holmgren, "The state of the soil is often the best measure for the health and well-being of society. There are many different techniques for looking after soil, but the best method to tell if soil is healthy is to see how much life exists there." Without healthy soil, producing an abundant crop is much more difficult. It is also important to value and respect the forests, grasslands, creeks, rivers, lakes, and oceans on the planet, and the multitude of life they

"Revolution is based on land. Land is the basis of all independence."

Malcolm X, November 10, 1963

support. Acknowledging the interconnectedness of life is a first step in employing permaculture ethics.

PEOPLE CARE

People care begins with taking care of—and taking responsibility for—yourself, and extending that care to family, friends, and community. As gardeners, we can grow food to feed ourselves and our families, but we can also share what we have, be it food, seeds, or skills.

A good example of people care is Ploughshares Nursery, which I helped start on the former Alameda Naval Station, a superfund site in California. Residents of the Alameda Point Collaborative (APC) helped build the infrastructure for the nursery, including fencing, irrigation, greenhouses, shade structures, nursery tables, and demonstration gardens. At the nursery, homeless youth and adults learn sustainable gardening techniques. Residents of the Collaborative also help to propagate the plants, and are involved in their care and sale.

A project of the APC called Growing Youth provides part-time employment for up to a dozen youth. The young people supply residents with low-cost fruits, vegetables, and honey every week. By meeting other people's needs in collaborative ways, the youth in turn become more self-reliant and the entire community prospers.

FAIR SHARE

I come from a long line of social activists and gardeners who instilled in me the importance of caring for the earth and for each other. To share willingly with others, we must feel we have enough. However, to create abundance it's not necessary to grow every possible crop. It makes more sense to grow what we can, and then find other gardeners to share and trade with. In this way, we all get a variety of food, even if we don't have the conditions or desire to grow every crop.

The Permaculture Design Course (PDC) I've been teaching for ten years at Merritt College, a community college in Oakland, California, illustrates the benefits of redistributing surplus. The Landscape Horticulture Department has an acre of gardens and orchards that the students have worked hard to design, build, and maintain. In exchange, they take fruits, vegetables, and extra plants home each week. This is affordable, accessible, and fair.

One of the benefits of growing fruits, vegetables, and flowers is that you can share your surplus yields with others. In permaculture, we call this fair share.

The Twelve Principles of Permaculture

One way in which permaculture differs from other methods of gardening is that it is not just a set of practical techniques; it is a way of thinking and of adapting to a particular ecology. Each garden, each family, and each community is different, so permaculture relies on observation and local knowledge. That's why, in addition to the underlying concepts of earth care, people care, and fair share, permaculture is built around twelve guiding principles. Whether you are starting a new garden, or introducing permaculture practices to an existing garden, these principles will help you to understand the design process.

1 Observe and interact

Permaculture relies on an understanding of your site and local conditions. Ideally, you should observe your site for a year in all seasons, learning the patterns of sun, wind, heavy rains, flooding, hail, snow, animals, noise, views, and the like. Even if this is not possible, do a thorough assessment of the site's intrinsic qualities and visit nearby gardens to see what grows well in your area.

2 Catch and store energy

There's a nursery rhyme about a squirrel collecting nuts during the summer to tide him over during the barren winter, and the permaculture principle of catching and storing energy echoes this lesson. There are many ways to catch and conserve resources when they are abundant so that you have access to them when they are unavailable. For instance, a greenhouse can catch and store the sun's energy to keep plants warm. With clever placement, a greenhouse can even provide passive solar heat for other buildings. Canning abundant summer produce for lean winter months is a way of storing food energy. Harvesting rainwater or recycling greywater from the house prevents valuable irrigation water from being lost to runoff or the sewage system, and provides water energy during dry months.

3 Obtain a yield

Of course the whole purpose of an edible garden is to yield crops. But there are other less tangible—but no less valuable—yields from a permaculture garden. A yield may be the exchange of skills or information from one gardener to another. Community gardens are good examples of this principle, where neighbors work together to mulch paths and build raised beds, tool sheds, fences, and trellises. School gardens are places for experienced gardeners to teach the next generation how to grow their own food. Elders can share their wisdom, young people can share their enthusiasm and energy, and people from different cultures can share seeds, plants, planting calendars, and growing techniques.

4 Apply self-regulation and respond to feedback

I always thought the Native American idea "think of seven generations" meant to think ahead seven generations into the future. But I have been shown that it also means thinking back to our own great-grandparents, grandparents, parents, and ourselves, as well as forward to our children, grandchildren, and great-grandchildren. In a garden, it means behaving as

TOP
One way to catch and store energy is to make sure that water does not simply run off your property, but is saved for use during dry periods. Rain barrels are one water harvesting tool that is easy to implement.

BOTTOM
My older daughter, Gitanjali, helps contribute to growing our food by delivering seedlings to their permanent homes in the garden.

though we are part of a continuum, starting with an appreciation of the harvest of the land stewards of the previous generations, and planting perennials and enriching the soil so that years later our future grandchildren can continue to enjoy and reap the harvest of our labors. Responding to feedback can also mean remediating our own mistakes or those of our predecessors. This may mean replanting unproductive areas of the garden, or improving soil that has been impoverished.

5 Use renewable resources

Trees are an example of a renewable multipurpose resource. From them, we get fruit, nuts, seeds, building materials, and fuel. They also provide shade during summer for cooling our homes, blocking the wind, filtering the air, and releasing oxygen. Fruit trees can yield crops for many decades and are a resource that connects us to our community when we practice the ethic of fair share. Even when the trees have finished their productive years, we can chop them down and use the wood to construct new beds, cultivate mushrooms, or chip them to create mulch, knowing that all decomposing wood will eventually be transformed back into soil.

6 Produce no waste

One of the great things about a permaculture garden is that there isn't any waste. Instead, we find ways to re-use the leftovers from our gardening efforts. Composting is one example, especially red worm composting, where creatures in the garden efficiently convert organic wastes like vegetable scraps into soil amendments that are then put back into the vegetable beds. The worms' digestive tracts convert food scraps into castings that enhance

You can use the cut logs from trees to build simple raised beds. This is an example of how you can practice the permaculture principles of producing no waste and using renewable resources.

HUGELKULTUR

Suppose you are clearing an area of weeds, removing unwanted ornamental plants, thinning trees, cutting back brambles, and getting rid of rotting logs. These materials are too unwieldy for the typical compost pile, and their woody structure means they will take years to break down. The permaculture way to reuse this waste material is the technique of *hugelkultur*. The term *hugelkultur* comes from the German word for mound, and the technique is fairly simple. You pile up all the materials and cover them with sod, tree trimmings, finished compost, and straw. Because the base is made of woody materials, *hugelkultur* beds retain a lot of moisture, and you can plant annual crops like tomatoes or squash right away. These hungry vegetables will thrive on the slowly decomposing pile.

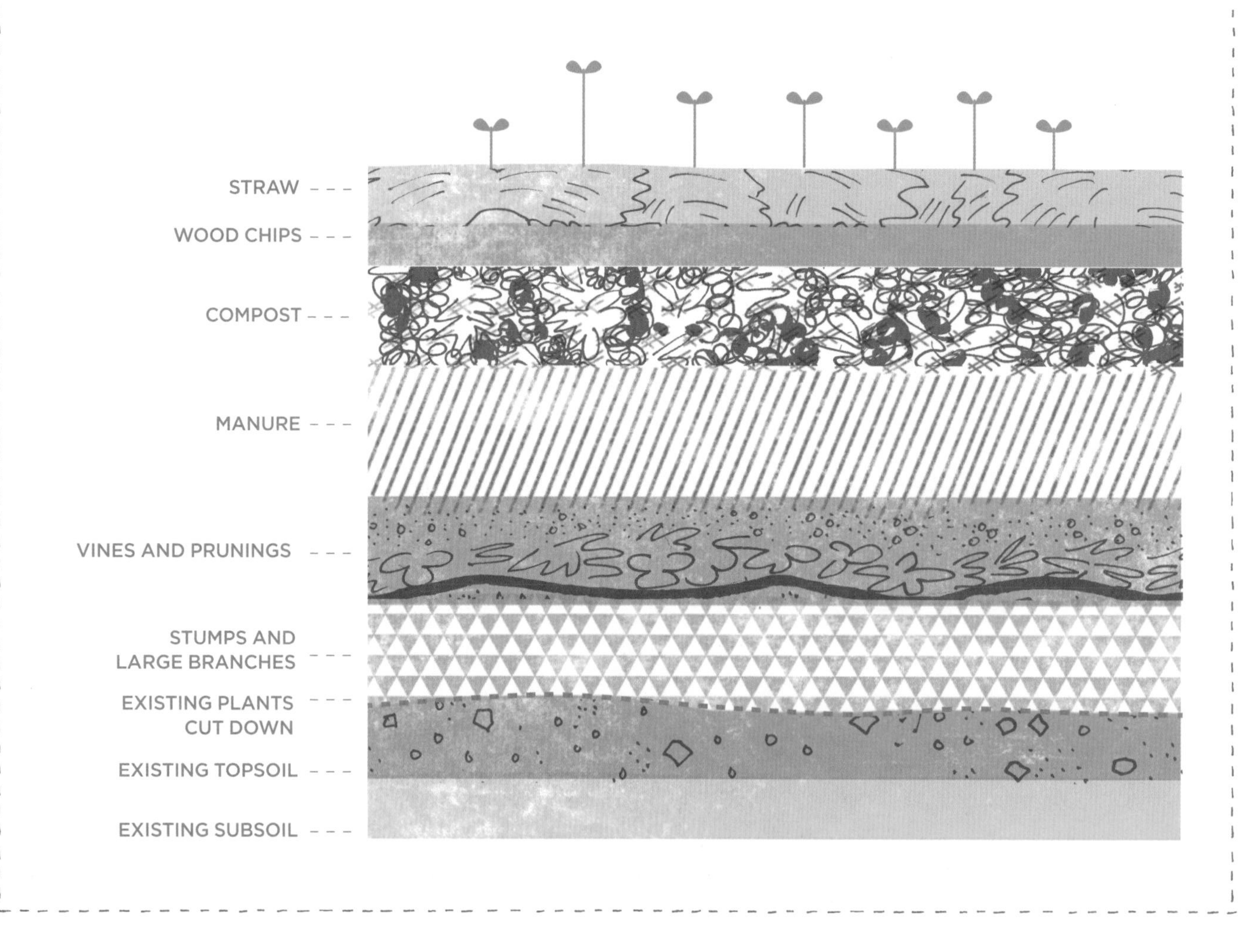

Christopher's Garden: INTEGRATION

When I moved to my current residence, the previous owners had built a garden using conventional modern landscape design that focused on a large lawn with several shade trees in the center and around the edges, raised vegetable gardens in the far back corner of the yard, and a perennial flower border separating the vegetables from the lawn. In the winter, the lawn flooded and the vegetables never saw any sun as the veggie patch was blocked by the neighbor's garage and fences.

After carefully observing, I was able to make productive changes. I sheet-mulched the lawn with woodchips, which solved the flooding, and replaced the ornamental trees and the lawn with vegetables and sixteen fruit trees. I turned the logs from the fallen trees into seven raised beds that are in the shape of flower petals, and added a central seating area and bamboo gazebo. Now the vegetables get more winter sun and are closer to the house so they are more accessible, and we have an abundance of fruit, flowers, and herbs.

My wife, Runa Basu, wanted a garden with flowers. I designed a central area of planting beds in the shape of a seven-petaled flower. This is how they looked after the ground had been sheet mulched and the beds filled with compost and mulch.

the soil food web and are the key ingredient for compost tea. This is a complete edible plant life cycle: from harvested crop, to kitchen trimmings, to the worm bin, and finally back to the garden as fertilizer.

7 Design from pattern to details

Permaculture seeks to understand and mimic successful patterns found in nature. For example, the spiral shape is found in everything from galaxies to the structure of DNA and the garden snail. It works well as a design template for an herb bed because it creates more surface space in a small area. A spiral-shaped bed also effectively creates microclimates because you can use some plants to shade others. This means you can grow sun-loving herbs like rosemary and thyme near shade-lovers such as mint and violets.

8 Integrate rather than segregate

Placing plants together in the right combinations helps them to grow in cooperation with each other rather than in competition. In this way, the whole garden

ecosystem becomes greater than the sum of its parts. And when you take the time to observe what is happening in an existing landscape, you can then find ways to make changes so that all the elements work to support each other.

9 Use small and slow solutions

In permaculture, we don't aim for the quick fix or the immediate payoff. The objective is to design a garden system that is composed of many small parts, each of which contributes in time to the overall function of the garden. An example is an emphasis on perennial crops. Perennials don't need to be replanted every year, so they save energy, and they don't disturb the soil like most annuals. Although their yields can be slower at first, perennials like chicory, dandelion, rhubarb, and sorrel produce earlier in cold climates because they are the first to come up in the spring. Similarly, permaculture focuses on small-scale, local solutions in preference to more industrial approaches. Sharing backyards, local produce swaps and giveaways, community gardens, and regional seed libraries are all examples of small and slow solutions.

10 Use and value diversity

Most gardeners love to look through plant catalogs for new varieties of vegetables to grow, and such diversity isn't just interesting, it's also smart. There is less vulnerability to a single disease or pest when different vegetables and varieties are planted in proximity, whether it's an entire farm or a backyard garden. During the Irish Potato Famine of 1845–1852, approximately one million people died and a similar number emigrated when a single,

Learning to select and save seeds from your favorite crops will save you money and ensure that you grow plants that meet your tastes and thrive in your conditions. These pepper seeds are easy to harvest and can be shared with others.

Seed Diversity

The Seed Savers Exchange (seedsavers.org) is a national, non-profit network of small growers and backyard gardeners who save heirloom, open-pollinated seeds. Every winter, I spend hours poring over their catalog, picking out some of my spring vegetables. To give you an idea of the diversity of their offerings, the catalog has more than one hundred pages devoted solely to tomato varieties. Each year I try something new, like fire-red 'Dragon' carrots or purple tomatillos.

widely grown variety of potato fell susceptible to a potato blight. In the Andes, where potatoes have been grown and developed for 5000 years, thousands of varieties are cultivated. Each year, a permaculture garden should feature some new varieties along with old favorites. This will build a diverse repertoire of plants and create a balanced garden system that can tolerate some losses without the entire garden failing. This helps to ensure resiliency in the face of climate change and other ecological challenges.

11 Use the edges

In a permaculture garden, we aim to make use of all possible space. This can mean designing vegetable, herb, and flowerbeds in unusual shapes. For instance, keyhole beds are modeled after an old-fashioned keyhole. Garden mandalas are circular arrangements of multiple keyhole beds. If you have six keyhole beds in a circle, one path will be the entrance and there will be a round area in the middle

Marginal spaces and structures can provide opportunities for growing. These Chinese bitter melons are growing over a wooden fence.

Garden Beds

Straight rows are a design of industrial agriculture to facilitate mechanical harvesting. An alternative in the home garden is to shape vegetable beds into beautiful curves that mimic nature and save space. Keyhole beds are modeled after an old-fashioned keyhole, with a narrow long entrance and a wider circular center.

Mandalas are geometric shapes that echo different spiritual traditions such as the Zen Buddhist sand mandalas. Garden mandalas are circular beds made up of four to eight keyhole beds. One space between keyholes becomes the entrance, and a circular area in the middle allows room to turn around. This clever design maximizes plantable space and minimizes path space. Orient the path to the south to get the most sun. Beds should be 3 to 5 feet wide, and the paths 2 feet wide. Plant lower-growing plants to the south and taller plants to the north. Lay out your plantings in terms of how often you will tend a plant—put the frequently harvested plants on the edge and the least-visited plants to the back.

Pathside, plant low-growing crops like strawberries and small cut-and-come-again greens like beet greens, leaf lettuce and salad mix, mache, and spinach. In the narrower middle beds, plant mid-sized vegetables like broccoli, cauliflower, chard, kale, leeks, and onions. Plant the largest crops on the outside, especially one-time harvests like cabbages, sunflowers, sunchokes, and yacon.

Save the big staples like corn, dried beans, fava beans, grains, potatoes, and squash for large, broad beds. Carrots, green beans, peas, tomatoes, and intensively harvested salad mixes can be planted in narrow beds for better access.

to give some room to turn around. This increases the number of edges to maximize plantable space and minimizes path space.

Marginal spaces that may not be suitable for traditional garden beds can also be turned into productive areas. Try growing heat-loving vines like beans, grapes, kiwis, melons, and squash on the side of a stucco or brick wall to benefit from the stored thermal heat and to soften the edge between the garden and the built environment. The vines also provide shade during the summer and let in light in the winter. Even dark nooks and crannies can be used to cultivate crops. I grow mushrooms under nursery tables, where they get ample water and little sun.

12 Creatively use and respond to change

Change is inevitable in the garden. What works well one season may not be successful the following year. Adapting to the shifting patterns of temperature, rainfall, pest populations, and other external forces is an important skill for the permaculture gardener. Our goal is to work with nature instead of trying to control it. As you face the challenges that come with growing edibles, keep this principle in mind. You'll soon realize that in the garden, there are no mistakes, just lessons pointing you toward better solutions.

Christopher's Garden: RESPONDING TO CHANGE

One change I've made in my yard is my composting system. For three years, I used black plastic stackable bins to make compost, but they were too small to sustain the heat necessary to activate the compost, and I once found a rat's nest while harvesting a pile. I replaced the compost bins with a salvaged 5-foot-square metal frame covered with ¼-inch hardware cloth. The structure is large enough to generate sufficient heat in the pile, and the wire keeps out the rodents. This compost evolution met my needs, addressed problems that came up, and turned into something more stable over time.

This one-acre garden in Sonoma County includes a food forest with fruit and nut trees, perennial and annual insect-attracting plants, mulched paths, and other permaculture essentials.

2 PERMACULTURE BASICS

The Polyculture Garden

FRUIT TREE GUILDS AND food forests are essential to permaculture. But what do these terms mean? In pre-industrial Europe, guilds were social groups organized around a particular craft or profession, such as blacksmithing or carpentry. In a permaculture guild, we organize edible plants around fruit or nut trees so that the plants are cooperative rather than competitive. Remember, in permaculture we integrate rather than segregate, combining elements in ways that make them interdependent. We are trying to design plant combinations that take a lot of the work out of the system for us as gardeners. Guilds cannot be considered native plantings, because they are not necessarily combinations found in nature. However, guilds are modeled on natural processes.

Polyculture means growing diverse plants together to create mutually beneficial relationships among them. The most famous polyculture model is the three sisters, a Native American polyculture of three plants. Corn grows tall and provides a support for beans, which act as nitrogen fixers for the soil, while squash covers the ground to suppress weeds and retain moisture.

In a monoculture like a banana plantation or a berry farm, a single crop is grown on hundreds of thousands of acres. Strawberry fields are even leveled with lasers for perfect flatness. On huge farms in the Midwest, miles and miles of what was once prairie is covered by a single species, whether wheat, corn, or soy. Chemicals are needed in the form of fertilizers, herbicides, and pesticides. The result is the consumption of lots of energy, both human and technological. Bill Mollison says that monocultures are maintained disorders.

LEFT
Polyculture plantings group plants together in ways that are cooperative rather than competitive. This grouping includes calendula, fava beans, thornless blackberries, tree collards, and tree tomato.

RIGHT
There are few straight lines in the permaculture garden. Fruit- and nut-bearing trees form a canopy over berry-bearing shrubs, flowers grow alongside vegetables, and herbs are tucked into containers and rock walls and at the edges of pathways.

In contrast, polyculture is a model from nature that offers more flowers and herbs, more beneficial insects, fewer pests, less maintenance, higher yields in less space, and more resilience to fluctuating temperatures and rainfall. Because there is so much abundance in your edible permaculture garden, the only problem becomes what to do with all the tomatoes.

Permaculturists grow as many different types of food and useful plants as possible—a version of not putting all your eggs in one basket. In my own garden, I have twelve

to fifteen species and varieties of vegetables growing in each bed. If some crops fail there are still other plants to provide food. Slugs may eat some of the lettuce or sunflower sprouts, but not much of the spicy mustard greens and tomatoes. If we only planted lettuce, the slugs may eat it all. By interplanting with vegetables the slugs won't eat, we guarantee some harvest for ourselves.

In the edible food forest, we assemble plants, animals, and fungi into guilds that function much like natural mixed-species ecosystems, with all the elements combining to create a productive and biodiverse polyculture.

25

Fruit Tree Guilds

The central element of a guild is a fruit or nut tree. You will usually find that it is easier to design guilds with deciduous trees because when the leaves fall, you can plant cool-season edible crops below—and most edible fruit and nut trees are deciduous. In larger gardens, evergreen trees can be used for other purposes, such as to provide windbreaks or create shade. Plants are organized around this central tree in supporting layers, each fulfilling a different purpose: fixing nitrogen, providing mulch, seeking nutrients, and attracting beneficial insects.

The beauty of tree guilds is that they do not repeat the same ingredients, but creatively combine elements to reflect local conditions. There is no one-size-fits-all garden, but the guild system will help you design a food forest that fits your individual needs and particular climate. A functioning food forest will perform the gardening tasks of adding nutrients to the soil, suppressing weeds, and attracting beneficial insects. Remember, the permaculture goal is always to make more food and less work.

FIRST SUPPORTER: NITROGEN FIXERS

The first supporters in the tree guild are plants that fix atmospheric nitrogen (N), a process performed through a symbiotic relationship between soil bacteria and the roots of certain plants. Simply put, without nitrogen, plants cannot grow. The air we breathe is more than 75 percent nitrogen and it is an essential nutrient for life—yet neither we nor any other animals or plants can directly absorb nitrogen from the atmosphere. However, nature has a solution. Certain plants from the legume family—most commonly peas and beans, but also many trees and shrubs—have evolved to provide carbon, water, and sugar for the bacteria that colonize their root systems. In exchange, the bacteria take nitrogen from the atmosphere and fix it, making it biologically available for the roots to assimilate. Plants use the nitrogen to produce proteins, enzymes, and amino acids that are consumed up the food chain by animals and humans. So planting these nitrogen-fixing legumes not only provides nitrogen to fuel plant growth, but it forms the basis of a vital system of nutrition.

Small trees and shrubs in the legume family are an ideal supporting layer for the guild. They provide extra nitrogen and they are perennial, so they do not need to be replanted each year. You can prune back or coppice (cut back hard each year) small trees and shrubby nitrogen fixers like acacias, agastache, alder, lupine, locust, mesquite, Siberian peashrub (*Caragana arborescens*), seaberry (*Hippophae rhamnoides*), tagasaste (*Chamaecytisus palmensis*). Leave all the clippings and cut branches in place as mulch. The trees respond to this pruning by dying back and releasing their stored nitrogen and organic matter.

SECOND SUPPORTER: LIVING MULCHES

Mulch is the next supporter in the fruit and nut tree guild. Essentially, mulch is a layer of material that covers the soil to a depth of several inches or more. Mulch suppresses weeds, retains moisture, and slowly builds soil fertility as the organic material decomposes. If you can generate your own mulch onsite, you will save both labor and money because there will be no need to import materials from an outside supplier.

This newly planted sideyard at the Boys and Girls Club in Petaluma, California, contains all the elements of a fruit tree guild. A young peach tree provides a crop, and a redbud is a nitrogen fixer in the shrub layer. Lemon balm, nasturtiums, and strawberries provide living mulches alongside a rainwater swale.

The large leaves of zucchini and other members of the cucumber family grow quickly to cover the ground, shading the soil and retaining moisture. Straw mulch performs the same function before the plants grow in.

Many types of mulch are familiar to gardeners, from straw to woodchips to sheets of plastic. But for a supportive layer in the tree guild, the ideal mulch comes from cut-and-come-again plants that regenerate throughout the growing season—a good example of producing no waste. Vigorous herbaceous perennials such as comfrey will send out new leaves after being trimmed multiple times in a season. I usually get twelve or more cuts a year from the comfrey in my garden; I tuck the trimmed leaves under mulch or stew them into liquid fertilizer. I also grow sunchokes (Jerusalem artichokes) as mulch plants. These giant perennial sunflowers will grow to 15 feet in height and are vigorous enough to be cut back a number of times in the growing season yet still make a sizable crop of edible tubers in the winter.

Annuals with large leaves, like pumpkins, squash, and zucchini, can also serve as living mulches, spreading quickly and shading the ground with their foliage. For smaller gardens, choose a compact bush variety like 'Table Queen' or your mulch will soon overrun the rest of your garden.

THIRD SUPPORTER: NUTRIENT CATCHERS

The next supporters in the fruit and nut tree guild are known as dynamic nutrient accumulators. While plants need the three main macronutrients—nitrogen (N), phosphorus (P), and potassium (K)—they also need micronutrients like calcium, iron, and magnesium. Nutrient catchers are plants with long taproots that forage deep in the ground to seek out these micronutrients and make them available to more shallow-rooted plants. Most accumulators specialize in particular nutrients, so they can be used to remedy a specific deficiency in the soil.

Many plants in this category are considered weeds, such as chicory, dandelion, and yellow dock. Yet not only do they provide nutrients for other members of the plant guild, but also you can eat them when they are tender spring greens, then keep cutting them back in summer to serve as mulch or to toss into the compost pile. Some common flowers are also nutrient accumulators, including borage, lupines, marigolds, and yarrow.

FOURTH SUPPORTER: INSECT ATTRACTORS

Insectary plants are the final supporter in the fruit and nut tree guild. Gardeners certainly have reason to be suspicious of insects in the vegetable garden, but entomologists estimate that 90 percent of common garden bugs are pollinators or pest predators and only 10 percent are likely to damage crops. Pollinating insects are essential for the production of fruit and seed crops, and predators can be our allies by reducing the number of pests that damage our plants. In fact, a food forest is a wildlife habitat that encourages a balanced insect population. This in turn attracts other insect predators such as

DYNAMIC ACCUMULATORS

Plant	Nutrient
alfalfa	N, Fe
arrowroot	Ca
borage	Si, K
buckwheat	K
burdock	Mn
calamus	N, P, Mn
caraway	P
chamomile	Ca, K
chickweed	K, P, Mn
chicory	Ca, K
chives	Na, Ca
clovers	N, P
comfrey	Si, N, Mg, Ca, K, Fe
dandelion	Na, Si, Mg, Ca, K, P, Fe, Cu
dock	Ca, K, P, Fe
fennel	Na, N, P
flax	K
grounsel	Fe
lamb's quarters	N, Ca, K, P, Mn
lemon balm	P
lupine	N, P
marigolds	P
meadowsweet	Na, S, Mg, Ca, P, Fe
mullein	S, Mg, K, Fe
mustards	S, P
nettles	Na, S, N, Ca, K, Fe, Cu
parsley	Mg, Ca, K, Fe
plantain	Si, S, Mg, Ca, Mn, Fe
purslane	Mg, K, Mn
salad burnet	Na, S, Mg, Ca, Fe
savory	K
sorrel	Na, Ca, P
sow thistle	Mg, K, Cu
tansy	K
toadflax	Mg, Ca, Fe
tobacco	N
valerian	Si
vetch	N, K, P, Cu
yarrow	N, K, P, Cu

Nutrient key

Ca (calcium)
Cu (copper)
Fe (iron)
K (potassium)
Mg (magnesium)
Mn (manganese)
N (nitrogen)
Na (sodium)
P (phosphorus)
S (sulfur)
Si (silicon)

birds and bats, which can eat problem pests like beetles, caterpillars, and mosquitoes.

Three main families—aster, carrot, and mustard—are insectary staples. If not harvested or pinched back, they produce copious small flowers whose pollen and nectar are accessible to the tiniest of parasitic wasps and flies. These beneficial insects may go on to devour problem pests, or use them as incubators for their offspring. Honorary mention also goes to the plants in the onion and mint families.

Food Forests

Food forests and edible forest gardens basically expand the concept of smaller fruit tree guilds. In a forest, there are layers of plants that start with a high canopy of trees, lower layers of smaller trees, shrubs, herbs and ground covers, and then plants that grow underground, the mushroom layer, as well as vines that clamber between the layers. In developing the basic principles of permaculture, David Holmgren realized that to make the best use of resources like soil, water, and sunlight, we must design from patterns in nature and stack plants in layers.

The tall tree layer is the foundation of the garden, just as a tree is the center of a guild. A typical arrangement is to plant tall trees on the north side of the garden and then plant everything smaller to the south in descending layers so that no layer shades the one below. Examples of good tall trees—20 feet and above—for this layer include edible nut trees like Korean pine (*Pinus koreana*), oaks, pecans, and walnuts, and fruit trees on standard (non-dwarfing) rootstock, like apples, pears, and plums. Squirrels may live in the tall tree layer and distribute seeds and nuts. Raptors like falcons, hawks, and owls may also perch in the branches and help to take care of rodent problems for you.

The small tree layer can contain fruit trees on semi-dwarf rootstock that grow smaller than standard trees—typically 12 to 20 feet. These can be apples, cherries, figs, peaches, pears, plums, and persimmons. For cold-winter areas, try coppicing (regularly cutting back to the ground) nitrogen fixers like alder (*Alnus* species). For mild winter

PAGES 44–45
Edible plants like (CLOCKWISE FROM TOP LEFT) *sorrel, lamb's quarters, chicory, borage, and burdock have deep roots that bring up nutrients from the soil, making them available for plants growing nearby.*

OPPOSITE
Edible plants fill in different layers in the food forest. Here, an almond tree (upper left) provides the small tree layer over collards, kale, and purple coneflower. A kiwi vine (upper right), which can grow to 20 feet or more, is part of the vine layer.

BENEFICIAL INSECT ATTRACTORS

The aster family (Asteraceae or Compositae) is the most diverse family of flowering plants (almost twenty thousand species), and includes artichokes, cardoon, chicory, dandelion, endive, lettuce, sunflowers, sunchokes, and yacon.

The mustard family (Brassicaceae) includes garden stalwarts like arugula, Asian greens, broccoli, cauliflower, collards, cress, kale, mustard, radish, and turnips. Some of these plants can get large (even to 6 feet) when you let them flower, so give them some room.

The carrot or parsley family (Apiaceae) includes carrots and parsley, of course, but also celery, celeriac, cilantro, dill, Florence fennel, and parsnip.

FOOD FOREST LAYERS

The structure of the food forest consists of eight main layers: the tall tree layer or canopy, the small tree layer, the shrub layer, the herb or herbaceous layer, the root layer, the ground cover layer, the vine layer, and the mushroom or fungi layer.

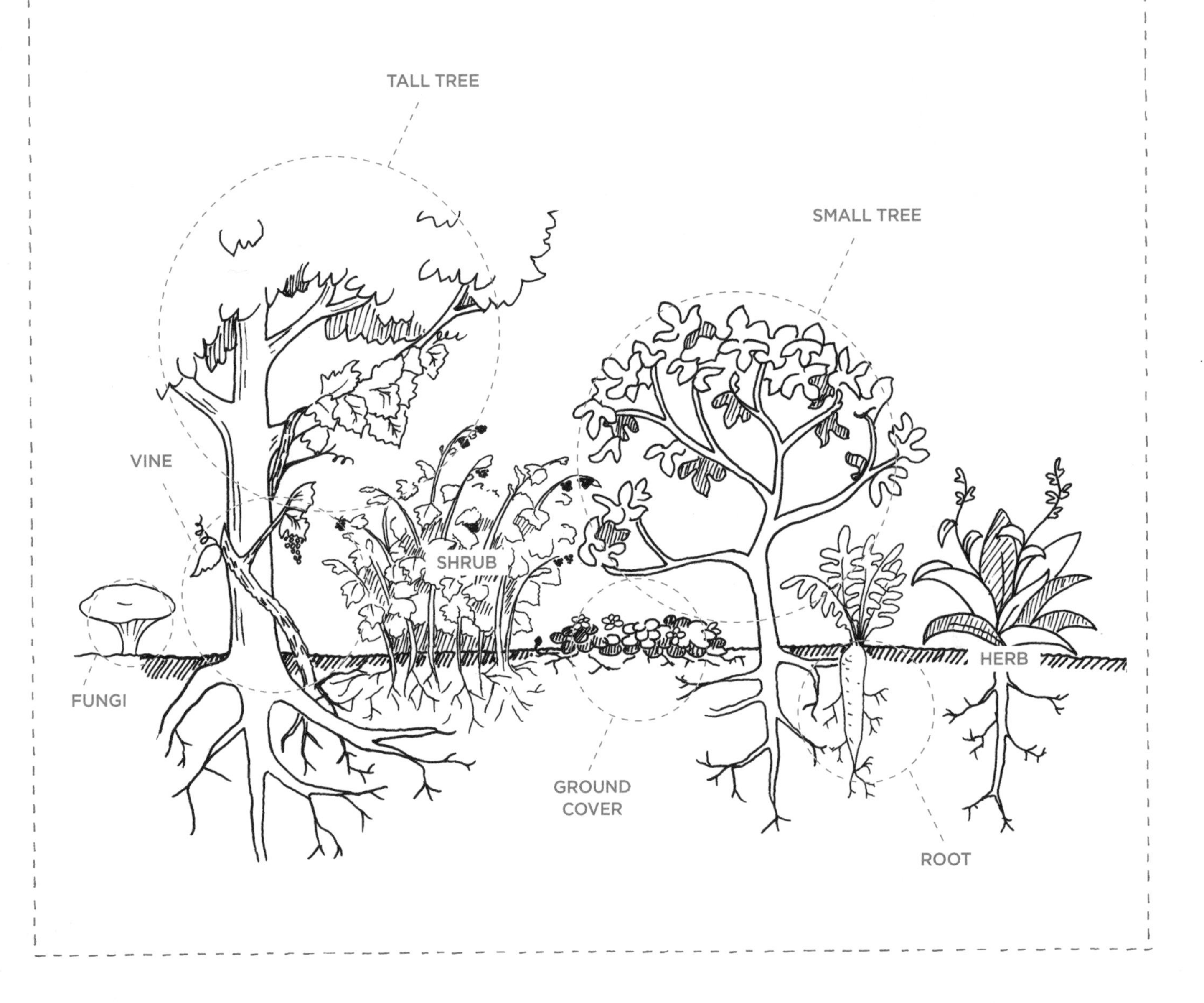

Bamboo

Bamboos are perennial woody grasses that can fit into several different layers of the food forest, depending on the species. Typically they are part of the shrub layer, but some can grow up to 50 feet in height, which clearly makes them part of the tall tree layer.

Bamboo is a true permaculture plant. Thanks to its rapid growth habit, it provides excellent and quick screening. Bamboo has a stately beauty, suffers few pests or diseases, and provides bird habitat. The strong culms, or poles, can be cut and used as building materials for everything from plant stakes to trellises, arbors, and fences. I have used bamboo poles to build a clothesline, a chicken coop, an outdoor shower, and an entry gate. The plants shed leaves that become mulch. Finally, the shoots are edible.

There are two basic categories of bamboo: runners and clumpers. The runners have given bamboo its reputation for invasiveness. Clumpers have a different kind of growth habit and don't overstep their bounds, but they are mainly suited to milder climates. You can avoid a lot of problems with bamboo by selecting the appropriate species and planting it correctly. My favorite runners include black bamboo (*Phyllostachys nigra* and *P. nigra* 'Henon') and for the coldest regions, nude sheath bamboo (*P. nuda*). My favorite clumpers include giant timber (*Bambusa oldhamii*), Buddha belly (*B. tuldoides*), Alfonse Karr (*B. multiplex*), and *Chusquea gigantea*, a solid core bamboo.

Bamboo needs good fertile soil, so add 4 to 5 inches of well-rotted horse manure and homemade compost to the planting area, and top that with another 6 to 12 inches of mulch. Runners need lots of space. In cold or dry climates, the conditions can keep them in check but elsewhere you will have to dig a trench and install a plastic rhizome barrier 30 to 36 inches deep. The trick is to not bury the barrier too deep beneath the compost and mulch. If you do, the bamboo can jump the barrier. I typically leave 2 inches of rhizome barrier above the mulch so I can spot any rhizomes attempting to escape, and cut them off.

Regular watering is necessary. For harvesting, dig into the mulch and compost around the base of the bamboo plants looking for the new young shoots, and cut them off with a sharp knife. Peel, and boil for twenty-five minutes and then stir-fry the boiled bamboo shoots. Different species will produce new shoots at different times of the year.

For use in garden construction, individual culms should mature for three to four years to harden. What is so amazing and sustainable about bamboo is that three years after it's established, you can harvest about one-third of the biomass annually without any appreciable detriment to the plant. Even if you have no use for the poles, thin out bamboo stands every year to maintain easy access around the grove.

areas, good nitrogen-fixing small trees that can be regularly coppiced include *Acacia* species, carob (*Ceratonia siliqua*), mimosa (*Albizia julibrissin*), and white lead tree (*Leucaena leucocephala*). Birds may nest in the small tree layer, eating insects that prey on other plants.

The shrub layer can contain fruiting plants like currants, highbush blueberry, elderberry, huckleberry, goumi (*Elaeagnus multiflora*), gooseberry, sea buckthorn (*Hippophae* species), and serviceberry (*Amelanchier* species). If you are in a cold-winter climate, remember hazelnuts and filberts (*Corylus* species) can tolerate a little shade and are great understory shrubs for nuts. Good nitrogen fixers for the shrub layer that are also highly ornamental include bush lupine (*Lupinus* species), California lilac (*Ceanothus* species), coffeeberry (*Rhamnus californica*), false indigo (*Amorpha fruticosa*), Siberian peashrub (*Caragana arborescens*), and tagasaste (*Chamaecytisus palmensis*).

The herb layer consists of low-growing herbaceous (non-woody) plants. This layer serves as living mulch to cool the soil and retain moisture. Think of a mix of annual and perennial vegetables, culinary and medicinal herbs, and edible and ornamental flowers that attract beneficial insects. In most areas, the herb layer dies back in winter. Asian greens, comfrey, daylilies, and salad greens are some of the more familiar members of the herb layer, but you can also try less well-known plants such as licorice milk vetch (*Astragalus glycyphyllos*), another nitrogen fixer.

The ground cover layer is made up of low-growing plants that spread horizontally. These can include fruiting plants like cranberries, creeping blueberries, kinnikinnick (*Arctostaphylos uva-ursi*), salal (*Gaultheria shallon*), strawberries, and wintergreen (*G. procumbens*). Mints are culinary herbs useful for making teas and flavorings as well as providing great bee fodder. In milder climates, creeping thyme, nasturtiums, and rosemary can cover large areas and make edible flowers and leaves. For nitrogen fixers, clovers make a great insect-attracting ground cover that can be mowed or left for grazing animals like ducks, rabbits, and sheep.

The root layer includes nutrient seekers with deep taproots like burdock (*Arctium lappa*, edible burdock, which can be found in seed catalogs under the Japanese name, gobo), carrots, dock, and parsnip. Other plants with fine and fibrous roots such as alfalfa, yarrow, and native bunch grasses also draw deep into the earth to pull out nutrients and moisture. When you cut back the root layer plants or they drop for the winter, the leaves rot and the stored minerals break down back into the soil.

Just like a tropical jungle, the food forest needs vertical climbers. Vines maximize small spaces, turning surfaces like fences, walls, and roofs into productive planting areas. Existing plants in the tree layer can act as supports for climbing vines, or you can build your own trellises out of prunings of bamboo, willow, or other woody plants. Vertical plants may be annuals such as beans, cucumbers, melons, nasturtiums, peas, squash, and tomatoes or

Encourage the vine layer by building trellises on which to train these scrambling plants. This trellis in my garden supports scarlet runner beans and mashua, an edible root nasturtium.

permanent perennial vines like blackberry, chayote, grapes, kiwi, mashua, Chinese yams (*Dioscorea batatas*), or passionfruit. A cold-weather native nitrogen-fixing vine that makes edible roots is American groundnut (*Apios americana*). Austrian winter pea, sweet pea, and vetch are non-edible nitrogen-fixing vines that can be planted as cover crops. The one caveat is to choose carefully as some vines are infamous for taking over in some climates (examples include kudzu, morning glory, and trumpet vine). Check with a knowledgeable local source if you are unsure.

These blocks of oyster mushroom mycelium are ready to be integrated into the food forest.

The last layer in the food forest is the mushroom, or fungi, layer. Don't overlook this layer; fungi are essential to any forest ecosystem. They need compost or sunlight to grow, and they convert dead organic matter like logs, paper, sawdust, straw, woodchips, and wood shavings into nutritious mushrooms. Just as in a natural forest, however, never eat mushrooms you have not positively identified. The safest are the ones you propagate yourself, including oyster mushrooms, which are easy to grow, shiitake, which are delicious and worth the wait it takes for them to mature, and the king stropharia mushroom (*Stropharia rugoso-annulata*), which can be grown in between other crops on hardwood chips, logs, or straw.

Putting the food forest layers together demonstrates the permaculture technique known as stacking, or increasing the yield of a space by integrating the components. David Holmgren calls on us to stack plants to make the best use of all our resources, like soil, sunlight, and water. Stacking may be physical, as in placing plants in order from the tree layer down to the mushroom layer. Or it may be based on timing, such as when we plant winter crops under deciduous fruit trees, knowing that the sun will penetrate through the bare branches to the herb layer below.

PLANT STACKING

Stacking may refer to plants or to a combination of plants, animals, and buildings. In this illustration, the largest plants form a high canopy—the tall tree layer. Then we step down through the other layers of the food forest into the soil itself, where much of the nutrient-creating work of the garden takes place.

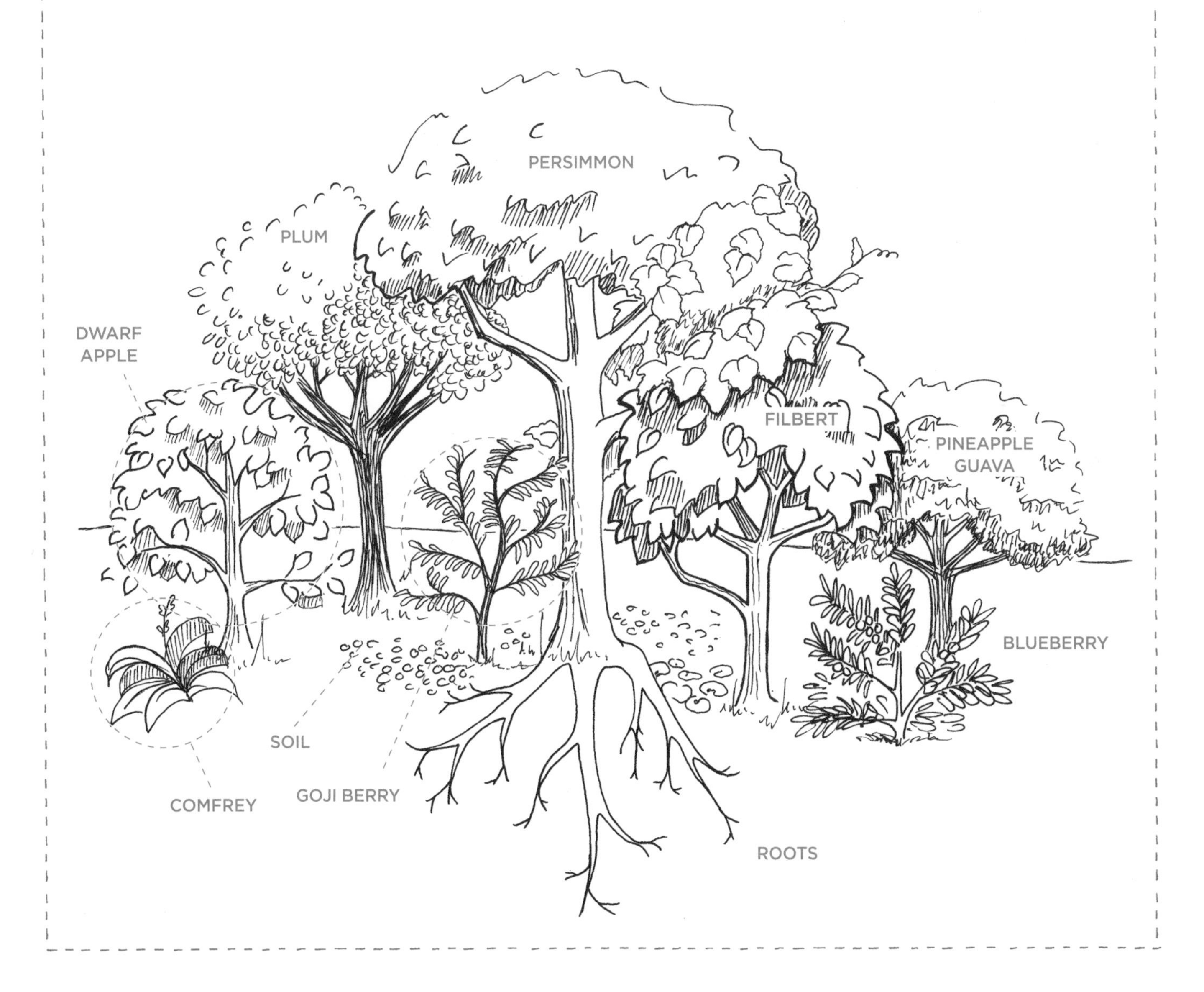

There are few straight lines in the permaculture garden. Curved beds and pathways maximize growing space and make tending and harvesting crops easily accessible.

3
DESIGNING
THE
PERMACULTURE
GARDEN

From Vision to Action

HOW WILL YOU TRANSLATE these somewhat lofty permaculture theories into your own garden? Before you can think about putting plants in the ground, the first step is to assess what is on and around your property and to learn about local gardening conditions. Then you can come up with an overall vision based on your location, your needs, and your understanding of the site. After that, you can develop a concept, and finally make a working plan for your garden.

Permaculture employs two special techniques for analyzing and designing your space: sectors and zones. These may sound like technical terms, but they are actually common-sense tools that help to organize your ideas and ultimately your garden vision.

Assessing Your Garden

Here are some of the elements that you should study when evaluating your garden space. Make as many (dated) notes as you can, take photos or videos, and use this information to inform your design and ultimate planting choices.

LOCATION

Your general location is the first starting point. All gardeners know that latitude determines the seasonal patterns of day length and hence the amount of sunlight the garden receives. The most important information you have when laying out the garden is the sun's path during the summer and winter solstices as well as the spring and fall equinoxes. In the desert, the sun can be

Learning about the energies that affect your garden, understanding your soil, and determining your needs will help you decide where to place planting beds, fruit tree guilds, and structures like chicken coops. Mandala beds (BOTTOM LEFT) *help to maximize planting space.*

overhead for much of the day, making it too hot to grow many edibles. In that case, creating shade may be your first priority. In more northerly areas, you must try to identify places where you can enhance the available sunlight by removing obstacles. You must also choose plants that can withstand short days and short seasons, as well as cold winter temperatures and spring and fall frosts. Perhaps you will need a greenhouse or cold frame for tender plants and seedlings.

PERMACULTURE IN COLD-WINTER CLIMATES

You can have a permaculture garden anywhere—even if your growing season lasts for one hundred and twenty days of the year and your winter temperatures drop down to single digits and below. Here are some design tips for cold-winter gardens:

- Choose short-season varieties.
- Grow root crops for winter storage.
- Capture and store heat with transparent materials like glass and plastic in the form of cloches, cold frames, hoop houses, and greenhouses.
- Cold climates often have poor soil because of the harsh weather conditions. Add plenty of organic matter and plant cover crops to improve the quality of the soil.
- Mulch often and heavily.
- Cold air flows downhill, so avoid planting at the base of slopes.
- If planting near the base of a slope, plant a frost trap—a triangle-shaped hedge with the point facing up the slope.
- Find the cold and hot spots on your property by planting tender indicator plants like nasturtium around the garden and seeing which ones do best in cold weather.

Hoop houses are simple to construct with a row of flexible PVC tubing or bamboo stakes. Clear plastic or row cover material helps to warm crops like these tender greens and shoots, extending the season in cold climates. Even placing row cover material over young crops can prevent frost and pest damage.

- Look for ways to trap heat for tender plants, such as planting against south-facing walls.
- Plant deciduous trees like quaking aspen around young tender plants. Aspens give a nice filtered shade and the fluttering of the leaves keeps cold air from settling during clear nights in late summer and fall.
- Use black plastic mulch to cover the soil and plant heat-loving crops like melons and eggplants through it.
- Build shelterbelts (barriers) on the north side of the garden.

Christopher's Garden: ASSESSMENT

I was able to visit and observe my current garden for an entire year because it belonged to my wife. During the period that we were dating, I noticed the seasonal flooding of the lawn, several potential growing areas blocked by shade trees, and existing vegetables struggling through half the year in the shade of fences and surrounding buildings. These kinds of observations can only take place through a well-informed reading of the landscape over several seasons.

- Collect leaves in fall and pile them around late-season crops like carrots, chard, collards, kale, and parsley.
- Observe your garden plants in winter. If certain specimens tolerate cold weather better than others, collect and store seed from those plants.
- If late or early frosts are expected, cover young plants with cardboard boxes, row cover material, or old bedsheets.

PERMACULTURE IN HOT CLIMATES

In hot, arid climates, plants face two challenges: too much sun and too little water. Here are some ways to make the most of gardening in these conditions.

- Look into zone 5, the wilderness, to find plants that thrive in desert conditions. They must be deep-rooted and able to withstand drought.
- Plant trees such as mesquite that can provide shelter for more familiar vegetables and fruit in the understory. This nitrogen-fixing tree has edible pods, and it attracts bees.
- Plant native hedges as windbreaks to lessen the drying effects of the wind.
- Look to indigenous cultures for traditional food crops that thrive in the area. Native SEED/Search is a seed company in Tucson, Arizona, that features low-desert and high-desert varieties.
- 'Hopi Greasy Head', 'Wekte', and other corn varieties have been developed by the Southwest Hopi people. Plant these corn varieties deep—up to 18 inches—so that the seeds have a natural mulch of soil above them.
- Cacti can yield several crops. The flower buds of the cholla cactus can be dried, pickled, or roasted and have a taste similar to asparagus. Prickly pears, or *tunas*, produce a delicious red or yellow fruit, and the

Starting seedlings in a greenhouse can help you get a head start on the growing season in areas where winters are long.

ABOVE
*Prickly pear cactus (*Opuntia *species) are native to the Southwest. The flesh of these desert fruits can be enjoyed fresh or squeezed into juice.*

OPPOSITE
From midsummer through September, mesquite pods ripen into a pale yellow-beige color. You can eat them as is, or dry them to mill into flour. Permaculture encourages us to look for such indigenous sources of food.

young pads harvested for *nopales*—a common ingredient in Mexican cuisine.

- Look to other desert cultures for ideas. Many traditional Middle Eastern gardeners plant date palm trees as a tall tree layer for shade. They fill in the understory with apricots, Chinese dates (jujube), figs, mulberries, olives, and peaches. Pomegranates and citrus fill in the shrub layer.
- Practice water harvesting and find ways to slow, spread, and sink seasonal water flow so that it is stored in the soil rather than running off your land.
- Plant with the beginning of the rainy season, and start with smaller plants so their roots can become established before the hot, dry season begins.
- Design your building wisely. Create shade with internal courtyards shaded by multistory buildings or by using shade cloth for single-story buildings. Cover an arbor with grapes so that you can plant herbs and fruit in pots below.
- The roof is the hottest place on the property, so use it for sun-drying tomatoes and other vegetables and fruit. Protect the drying food from birds and insects by covering it with window screens or floating row covers.
- Create cool, moist microclimates in courtyards by planting trees, building water features like ponds, and mulching the ground heavily.
- Build sunken beds instead of raised beds for maximum water retention. Plant trees in mulched pits or sheet mulch around the young plant (sheet mulching involves covering the ground with layers of organic matter). Keep the mulch away from direct contact with the trunk by stacking rocks around the trunk.
- Irrigate at night or dawn so that water doesn't evaporate, and use drip irrigation rather than sprinklers.

THE PATH OF THE SUN

Knowing where the sun traces its path across the sky is important for both winter and summer gardening. Observe whether there are any structures or large trees blocking the sun from reaching areas where you hope to grow edibles. In general, in the Northern Hemisphere you will plant the tallest layers of the food forest to the north and step down the lower layers to the south.

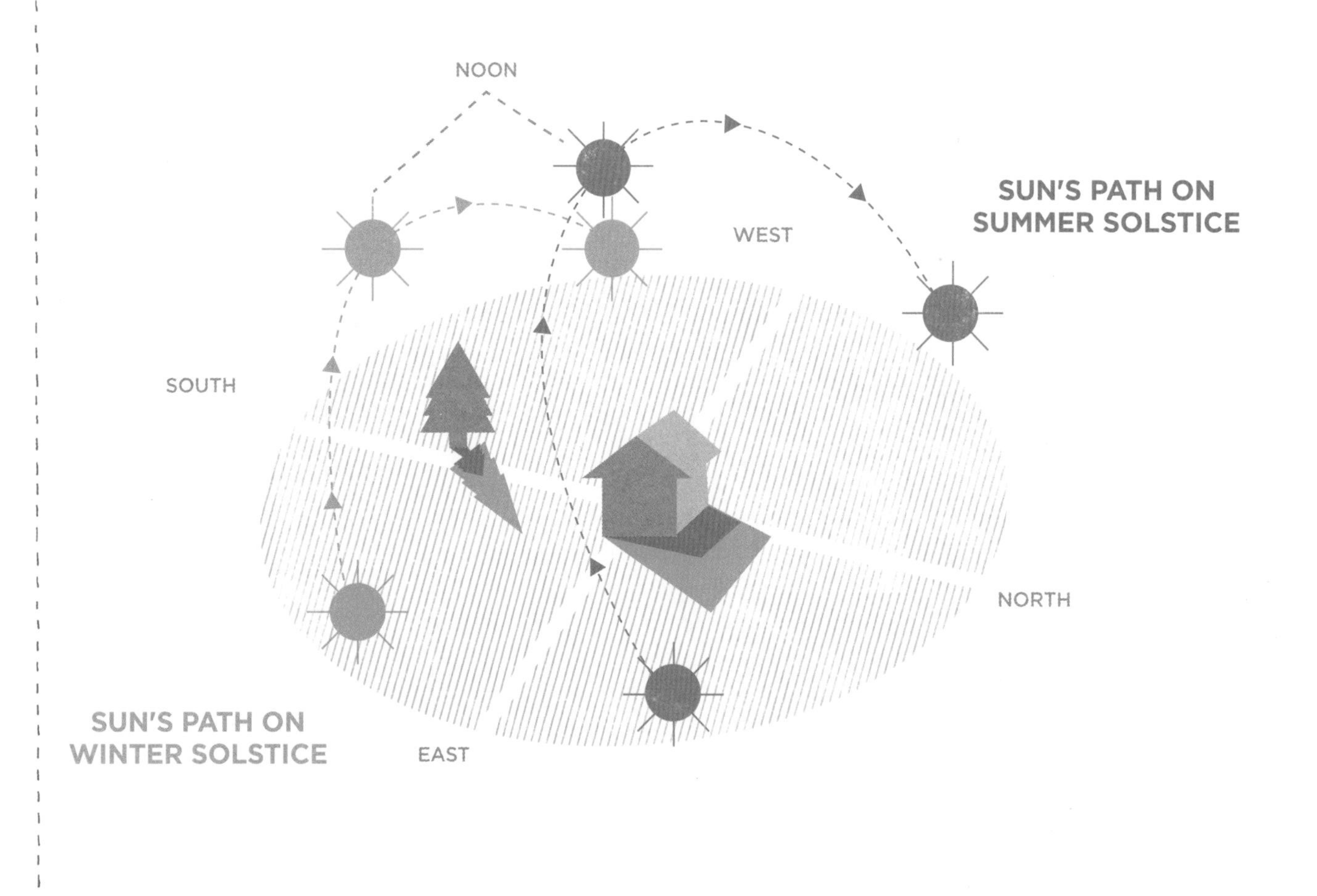

SOIL

Soil is the basis of gardening, and you cannot grow good food without good soil. But no matter how poor your soil, it can be made better. To understand how best to improve your soil, you need to know what you are dealing with.

You will probably be able to tell if your soil broadly fits into the categories of sandy, clay, or loam. To learn precisely what kind of soil you have, get a soil test from a commercial laboratory. This will typically tell you the percentage of mineral nutrients, pH, and whether any contaminants are present. At the very least, you can purchase inexpensive pH strips to determine whether your soil is acid or alkaline. Composting, mulching, and planting nitrogen fixers and nutrient accumulators can help adjust most minor mineral deficiencies. If your soil is severely deficient, you may need amendments; a soil test will give you a good idea if this is necessary.

Regional Soils

Soils are incredibly diverse, even within the same city and county. Where I live in the San Francisco Bay Area, most people have heavy clay soil. In Alameda however, just twenty minutes from my home, the soil is sandy and silty.

Most cities and counties have local soil conservation districts, which are a good source for soil information. However, it is wise to also do some on-the-ground investigating because land use patterns can change soils from yard to yard. If possible, find out what was on your site before any homes were built, whether your house was built on fill, whether the site was graded and compacted, and what the previous owners and tenants used the land for.

EXISTING PLANTS

Pay attention to the placement and size of existing trees and how they may have an impact on your garden in the coming years. For example, a coniferous evergreen or a fast-growing eucalyptus may soon get so big it shades the entire yard, making a sun-loving edible garden impossible. Evaluate all trees to determine what to keep and what to remove. If possible, it's better to get rid of potentially problematic trees when they are small. Call in an arborist for a second opinion on any major tree that is in question. A qualified arborist can determine the age and health of your larger trees, and help in locating someone to mill up the wood. In any case, keep the tree trimmings for mulch and the logs for firewood, mushroom logs, or raised beds. If you have smaller ornamental plants that will not fit into your garden vision, consider digging them out and donating them to a local school garden or plant swap.

Bear in mind the principle of valuing diversity and look to local native plant communities for insights in how to recreate ecosystems in the garden. What are the native plants in your immediate area? How have they adapted to the climate and how do they change with the seasons? What are the edible native plants in your area? Bill Mollison says to start first with native plant foods and then move to exotics. Native plants are almost always low-maintenance, they attract local pollinators, and they can be planted in difficult areas where other plants will not thrive. I have sun-loving natives planted in the dry curb strip and shade-loving natives planted on the north side of my house as a pollinator shelterbelt that also buffers the street noise.

Sectors

Sectors are a functional tool for understanding the garden's relationship to the surrounding environment. Broadly defined, sectors are external forces (or energies) that will affect your garden, such as sun, wind, rain, or even a loud or inquisitive neighbor. Some wild energies are welcome, like winter sunlight and cooling summer breezes. Others, like floods, cold winter winds, ocean spray, fog, or fire hazards may call for preventive landscaping. Are you on the coast or by a body of water? This can mean your garden is subject to moderating maritime influences or salt-laden winds, depending on the prevailing weather patterns. If you are inland, are there any seasonal weather events like cold winter winds? Are there mountains to provide a rain shadow or to trap fog? Are there nearby wetlands, wild lands, or animal habitats that may bring pollinators or pests to your garden? Are there local recommendations to provide a buffer zone to protect against potential brush fires?

Existing features like nearby evergreen trees, buildings, and slopes can all be incorporated into your final design, so mark them on your initial assessments of the property.

Now take time to reflect on your site, using sectors as a frame of reference. How big is it? Is it a standard suburban lot or does it have any irregular boundaries? What is happening in and around the site? Are there areas where water drains into your garden? Where are existing pathways or driveways that are heavily used? Look beyond your own garden, to your neighbors' properties. How close is your boundary line? Do they have trees or buildings that could shade your garden? Are there any vacant lots close to your property that could be a source of wind-borne weeds or perhaps even beneficial insects? How about the neighbors' views of your garden, and your view of theirs? Are there any zoning issues or setbacks that you will need to consider?

The elevations in your garden are also important. Look for low spots that could potentially be areas of poor drainage. Frost moves downhill and also settles in low-lying areas. On the other hand, south- and west-facing slopes are warmer hot spots. Are there any steep slopes where erosion might be a problem?

You can sketch up a map that includes the physical landscape: buildings, paths, roads, trees and other plants, rocks, slopes, gullies, and watering sources. Overlay that with various sectors drawn on tracing paper. You can start to see how the sectors affect the garden and each other.

SECTORS

Mapping sectors on your property can help you understand the external elements or wild energies that will affect your garden design.

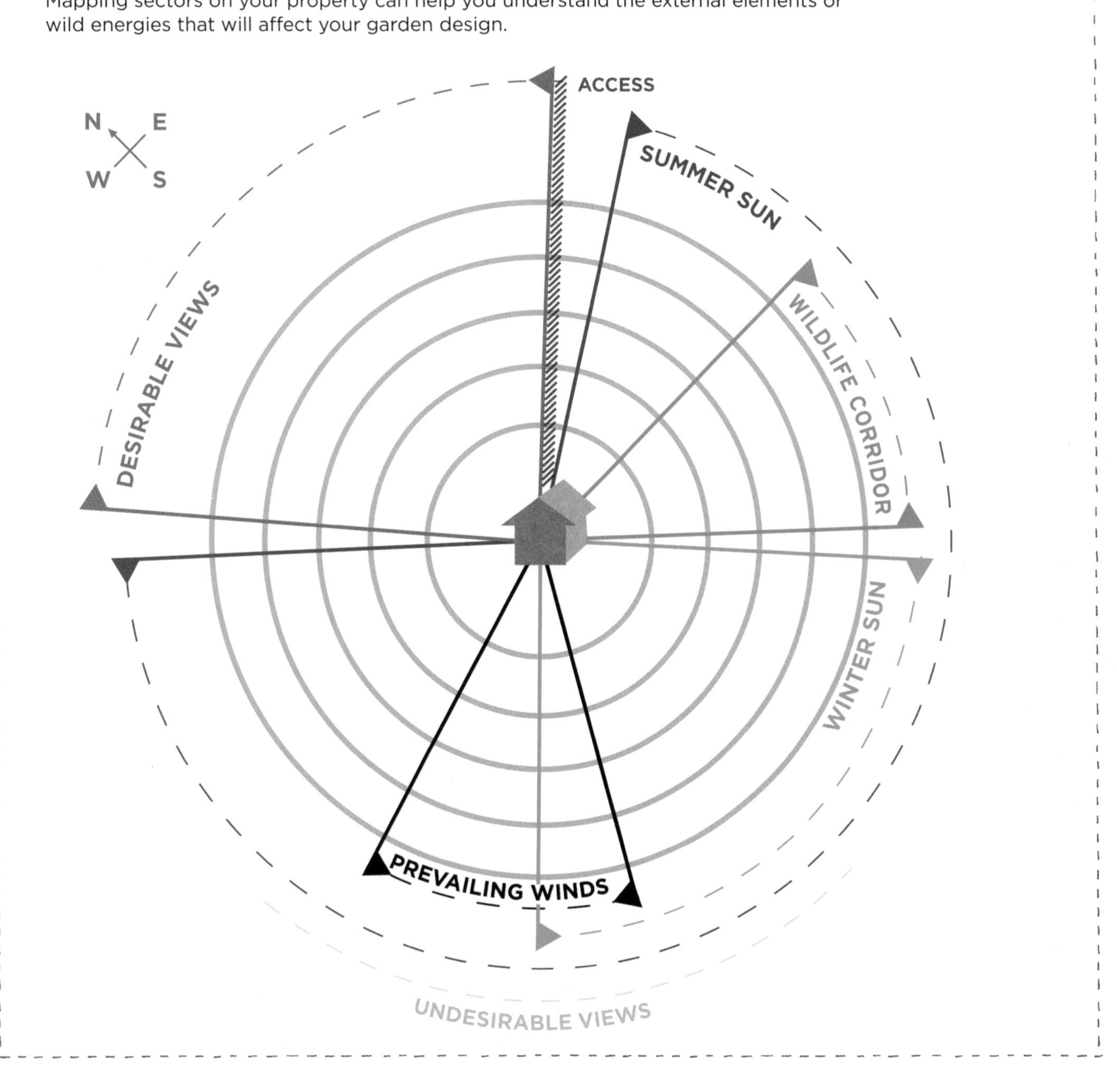

RAISED LOG BED

Raised beds can solve minor drainage problems in the garden. Even a height of 6 to 12 inches will allow water to drain more freely away from your plants' roots. If you have logs from cutting down a tree, you can set them in place to enclose a planting bed. Fill the inside with soil and generous amounts of compost. One advantage of using logs is that you can shape the bed to any proportions, thus maximizing space in the garden.

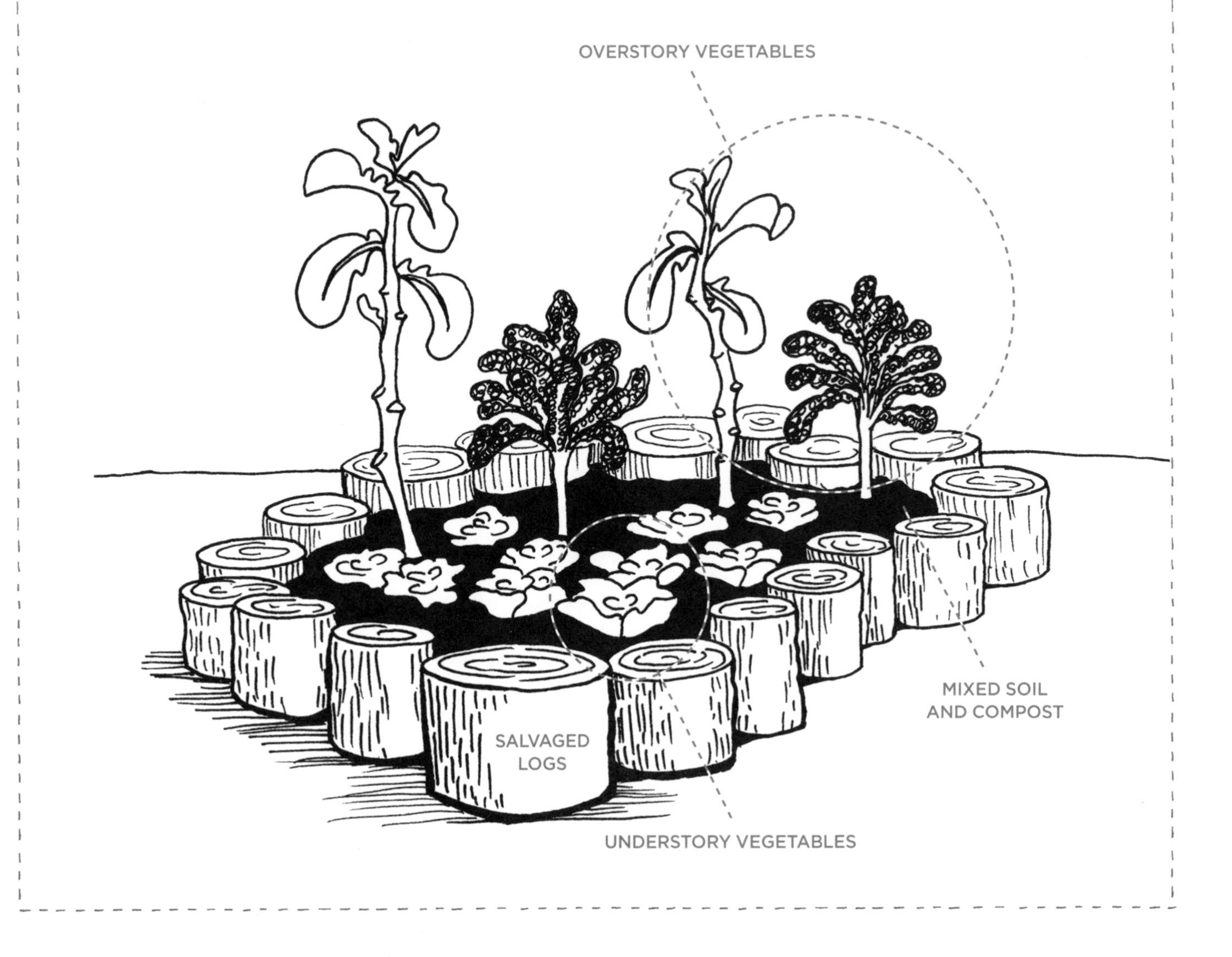

Water in the Garden

Water is one of the main aspects of any garden system. First you need to know how well your soil drains, which depends on the type of soil. Clay soil drains much more slowly than sandy soil. You can evaluate how water drains through your soil with a percolation test, which simulates months of rain or a really torrential downpour. Dig a few 1-foot holes around the potential garden site, fill them with water, and then time how long it takes for the water to drain out of the hole. If the hole drains quickly, you're in luck. If it takes an hour, you have clay or compacted soil that will probably need a lot of compost and mulch. Raised beds can help to alleviate poorly drained soil. More serious drainage problems may call for water infiltration or other techniques.

Water can be funneled from the roof into barrels and stored there for future use in the garden. This rainwater capture system is in Christopher's backyard.

PAGES 70–71
Water infiltration swales capture and retain water. They can be constructed in any garden, whether on a rural property or streetside in an urban setting.

WATER HARVESTING

Note where there is any running water in the garden, such as brooks, streams, or creeks. Are there underwater springs on the property? If you are able to observe your garden over a period of time, you will note any seasonal natural water features like a dry creek or a vernal pool. If you can, go outside during a heavy rainfall and see if there are spots where water collects or runs off the property.

Roof catchments are one way to harvest water. Even in low-rainfall areas, you can lose hundreds of gallons of rainwater that could be captured and used during dry periods. The water can be channeled through downspouts into cisterns, rain barrels, tanks, or into onsite swales and pools.

Greywater can also be harvested from the house interior, drained from baths, showers, washing machines, and sinks—any household wastewater excluding toilet wastes. Greywater systems vary from simple, low-cost solutions to highly complex and expensive systems. Sophisticated systems treat greywater prior to disposal, using settling tanks and sand filters to remove solids and pathogens. Before installing any greywater system, check with your local building department to see if there are any local codes regulating greywater. Always be sure you have a valve that returns water to the sewer system in case you have to make any repairs to the system.

Conventional drainage practices seek to move stormwater off the property as quickly as possible. This can cause soil erosion and carry sediment through sewage systems into streams and lakes. An alternate approach is

Building A Swale

Swales are dug on contour with level bottoms so that the water percolates evenly through the base of the swale. To position the swale correctly on a slope, you need to find the contour line, an (imaginary) line that is the same level across the slope. If the swale is positioned exactly along the contour, it will most effectively stop the flow of water. Soil excavated from the swale is mounded up on the downhill side of the ditch. This berm can be planted with permanent plants to stabilize the soil.

You cannot guess at contour lines; you must measure them. A useful home-made tool for marking out contour is a weighted A-frame level. This consists of three poles fastened together in the shape of a capital A. Take a small rock and tie some string around it, then tie the other end of the string to the top of the A-frame so that the weight swings freely below the level of the crosspiece. Place the frame on level ground and mark the place where the string touches the crossbar. This is your level mark. Use the A-frame to find points across the slope that are the same level and mark them with flags.

Instructions

1. Mark out the contour line on a 3:1 slope or less with flags and A-frame.
2. Dig the trench approximately 3 feet wide and 1 foot deep, mounding the excavated soil to the downward slope to form a berm.
3. When you think the trench is level, then place a garden hose in the bottom of the swale and flood it. Water will seek its level and you will be clearly shown where it's low and high. This will help you to finish mounding the soil downhill.
4. Fill the trench with 6 inches of tree trimmings.
5. Pile compost on top of the berm.
6. Plant the berm with trees, shrubs, vines, and cover crops.
7. Mulch the entire trench and berm with a further 2-inch layer of straw, a 1-inch layer of sawdust, or 4 inches of tree trimmings.

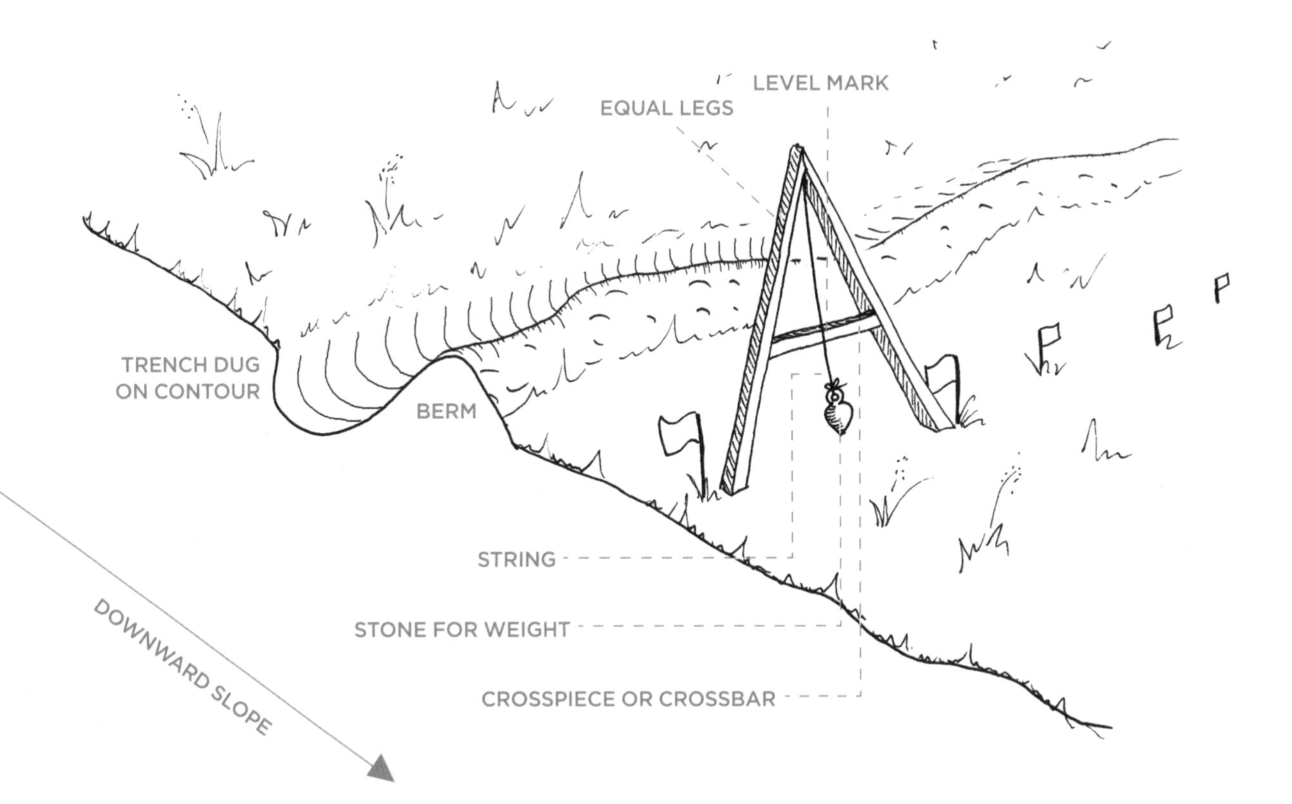

to slow down and capture the runoff by various water infiltration techniques, which may include earth-forming basins, berms, dry creeks, pools, and water infiltration swales. The objective is to channel water along the longest path possible so you slow, spread, and sink the runoff around the property. The soil acts like a sponge and stores water in the ground for plants to use later in the season.

In arid climates, sunken beds can be used to collect rainwater. Dig out the garden beds to a depth of 6 inches. Fill the sunken pit with soil and compost, plant, and mulch.

TRENCH COMPOSTING

A variation on swales is trench composting, an indigenous hillside farming practice from Central America. Swales are laid out on contour, but the trench is dug lower and wider than a regular swale. The bottom 6 inches of the trench is filled with organic matter like kitchen scraps, coffee grounds, or horse manure. These reserves of organic material will decompose over the year into finished compost with no turning.

The trench is then capped with woodchips, straw, or some other mulch material. The following year you can dig out the trench and move it slightly downhill to the berm of the swale, mulching the existing perennials and trees. I find these trenches can do double-duty and make a fine network of paths for moving about the garden.

Deadwood swales are another strategy to capture water. After marking the swales with the A-frame and flags, you dig a shallow trench and place logs where the berm is to be built. As you proceed with the swale construction, bury the logs with the soil dug out from the trench. Over time the wood will rot and act as a sponge to hold water. The buried wood is also a good way to sequester carbon in the soil.

Drip irrigation is the most effective method of providing plants with consistent watering during periods of drought. Tubing with emitters (OPPOSITE TOP) *and soaker hoses* (LEFT) *release water slowly so there is little evaporation.*

Designing Your Garden

Consider the day-to-day use of your garden. Where you want to socialize? Do you want to have herbs growing near the kitchen? Will you put in raised beds? Will you have a shed, chicken coop, goat enclosure, greywater system, beehive, or other structures even if you're not ready to build them yet? To help determine the best placement for plantings and structures, permaculture offers you a way to look at the garden in terms of zones.

PERMACULTURE ZONES

Learning how to identify the six permaculture zones is the best design tool for creating a basic gardening layout. Think of a small stone thrown into water and the resulting ripples in concentric circles. Permaculture zones function similarly, although the boundaries within zones are not as clear cut as circles, but instead overlap depending on existing site features. Planning in terms of zones helps you determine where in the garden to place elements based on their frequency of use and their scale. Most average-sized home gardens have mainly zones 0, 1, and 2, but with a little creativity, we can envision the other zones even in small residential gardens.

The first zone is numbered 0, and it represents the home and the self. Permaculture designer and teacher Bonita Ford says, "Zone 0 reminds us to begin the design process with ourselves, with observation of the inner landscape. Our personal values, needs, likes, and dislikes all contribute to the design and its implementation. The invisible/social aspects of our design ask that we sustain our own vitality and creativity, taking care of ourselves as living elements in the design." Clearly, if we are not doing enough self-care, then we will have a hard time tending a garden. Zone 1 starts outside the back door, or near another highly used area, and zone 5 is the farthest away from the living area and so the least used. It represents wilderness. We may need to plant here to restore native habitat or to attract pollinators.

We want to have the most tender and most frequently picked edibles closer to home. A sample annual planting plan might be:

- **ZONE 1:** herb garden and salad mixes
- **ZONE 2:** frequently picked collards, kale, and Swiss chard
- **ZONE 3:** tomatoes or potatoes
- **ZONE 4:** amaranth, fava beans, mushrooms, sunflowers, and winter squash
- **ZONE 5:** native plants

STRUCTURES

Zones can also help you decide where to put structures (sheds, greenhouses, arbors, fences, and gates) and infrastructure (irrigation and lighting) in the garden. Remember that you want to place elements according to energy use, frequency of care, and quantity of input and output. Although you can keep some elements like rain barrels in several different zones, a sample plan for garden structures might be:

- **ZONE 1:** seating and dining areas, small greenhouse, cold frames, trellis, patio, compost bin, worm compost bin, nursery table, rain barrel
- **ZONE 2:** large greenhouse, tool shed, wood storage, chicken coop, beehives, well, swales, greywater
- **ZONE 3:** windbreaks, firebreaks, small ponds, workshop, storage, drip irrigation

PERMACULTURE ZONES

ZONE 0 **The Home and the Self:** most important
ZONE 1 **Doorstep:** intensively cultivated and cared for
ZONE 2 **Garden:** semi-intensive cultivation and care
ZONE 3 **Farm:** semi-frequent three-season cultivation and care
ZONE 4 **Semi-wild Woodland and Food Forest:** minimal care
ZONE 5 **Wild Conservation Land:** unmanaged nature

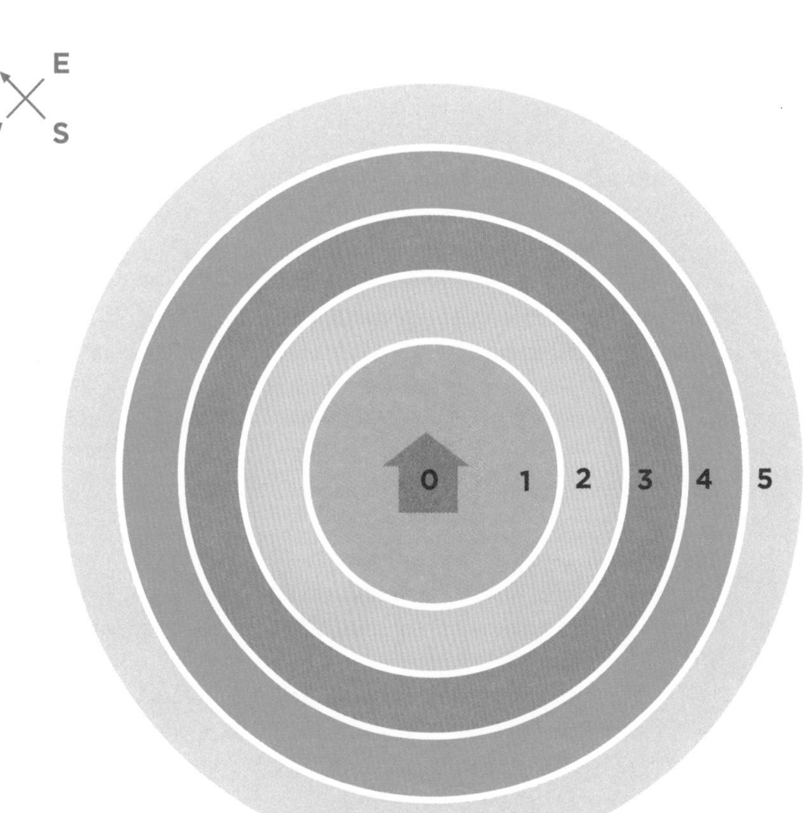

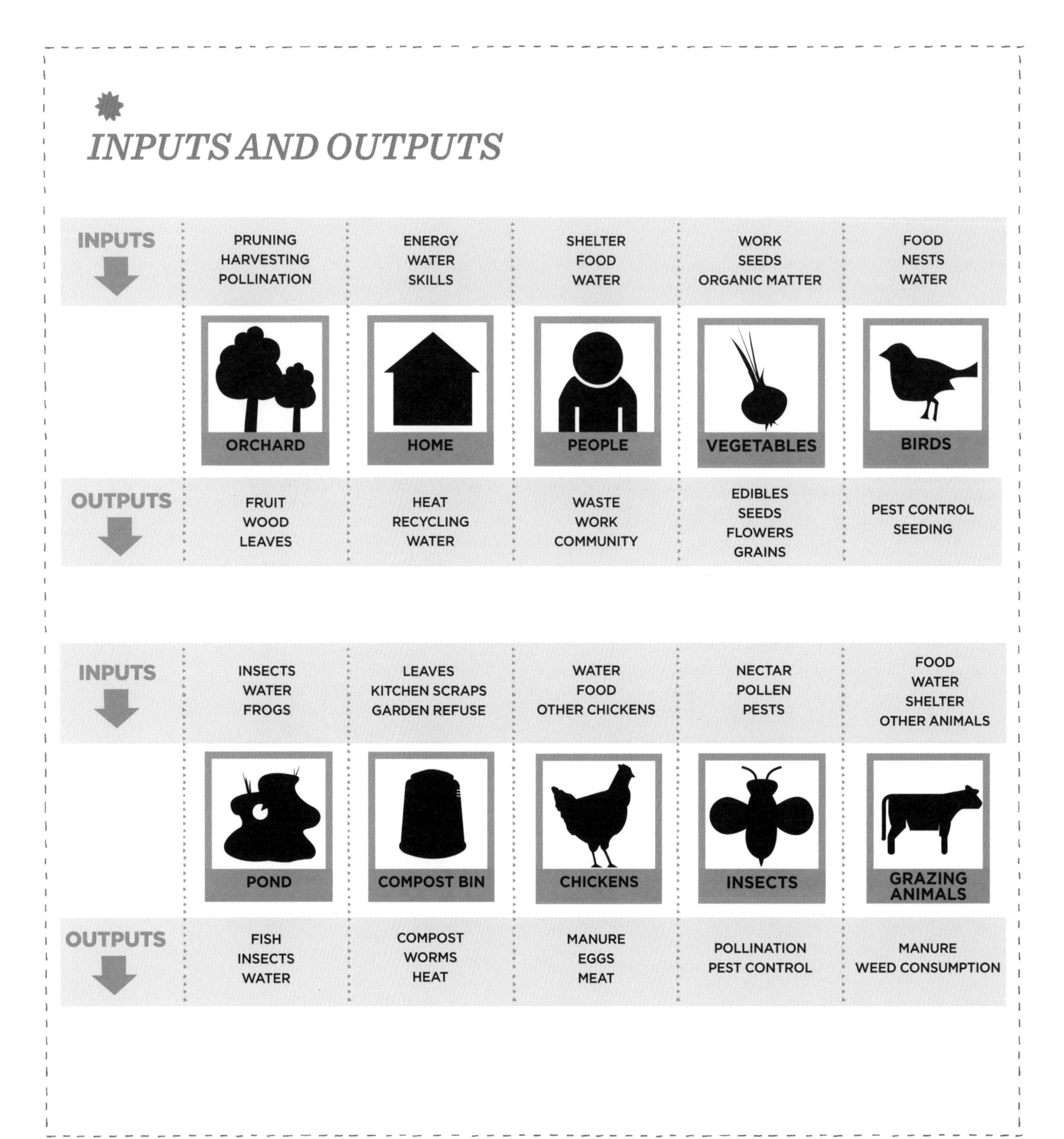
INPUTS AND OUTPUTS
INPUTS
PRUNING
HARVESTING
POLLINATION
ENERGY
WATER
SKILLS
SHELTER
FOOD
WATER
WORK
SEEDS
ORGANIC MATTER
FOOD
NESTS
WATER
ORCHARD
HOME
PEOPLE
VEGETABLES
BIRDS
OUTPUTS
FRUIT
WOOD
LEAVES
HEAT
RECYCLING
WATER
WASTE
WORK
COMMUNITY
EDIBLES
SEEDS
FLOWERS
GRAINS
PEST CONTROL
SEEDING
INPUTS
INSECTS
WATER
FROGS
LEAVES
KITCHEN SCRAPS
GARDEN REFUSE
WATER
FOOD
OTHER CHICKENS
NECTAR
POLLEN
PESTS
FOOD
WATER
SHELTER
OTHER ANIMALS
POND
COMPOST BIN
CHICKENS
INSECTS
GRAZING ANIMALS
OUTPUTS
FISH
INSECTS
WATER
COMPOST
WORMS
HEAT
MANURE
EGGS
MEAT
POLLINATION
PEST CONTROL
MANURE
WEED CONSUMPTION

- **ZONE 4:** swales, wood lot, nut trees, barn, pasture, large ponds
- **ZONE 5:** native habitat, wildlife corridors, river, wild berries, wetlands

INPUTS AND OUTPUTS

The natural world is composed of complementary pairings where individuals benefit from proximity to each other. The waste created by one entity can provide nutrition for another. Compost is a perfect example of this symbiotic relationship, where inedible waste is turned into fuel for edible plants. As stated by Bill Mollison and Reny Mia Slay (also a permaculture pioneer), "We set up working relationships between each element, so that the needs of one element are filled by the yields of another element."

To produce no waste, you need to create closed systems that link the garden's inputs (needs) and outputs (products or yields). The most important linkage for us, of course, is that a vegetable garden's output meets the need of food for our family and community. But there are many other closed systems you can create within the permaculture garden.

For example, the home's wastewater output (greywater) can be redirected to irrigate fruit trees and bamboo. Similarly, urine can be diluted and used to water the vegetable garden and orchard. When you harvest edibles and flowers, you can compost the spent plants, skins, cores, stems, and leaves, or feed them to the chickens, ducks, or goats. Poultry can peck their way through the garden to eat slugs and other pests, leaving nitrogenous manure that feeds the soil. You can build trellises for bird habitats, and leave logs to encourage mushrooms. You can allow lettuces and other annual vegetables to bolt, and then collect the seeds for next season's salad garden.

In the spirit of people care, yields are also gained through positive interactions with the community and the encouragement of garden practices that reap more fertility and less work over time.

Christopher's Garden: VISION

My overall vision was education, edible gardening, compost generation, water harvesting, privacy, creating a sanctuary, and raising a few chickens and ducks. I wanted a special place to introduce my young family to the natural world, and to create a teaching garden for students. My vision included a food forest—a healthy mix of fruit and nut trees with flowers, herbs, and vegetables working together to maximize diversity and yields. I also imagined a little studio for working on other garden designs. Blocking out neighboring buildings that were being remodeled into condos would provide privacy. With those specific goals in mind, I was more easily able to create a concept and action plan for the garden.

Goats, sheep, and pigs are excellent land-clearers, grazing their way through almost any type of vegetation. If you have a large amount of land to clear, consider borrowing some of these animals for natural weed control. As a bonus, they will fertilize the soil for free.

VISION

Now you are ready to let your imagination run wild in the design process. What would you do with your garden if there were no limits? Start by brainstorming: what are your goals, desires, needs, themes, and priorities? What would benefit the garden? Do you want to grow culinary herbs for the kitchen, medicinal herbs for tea, fresh fruit for picking, tomatoes for canning, grapevines to produce your own wine, or flowers for cutting? You'll have plenty of opportunities later to prune back your dream design, but for now, let it flourish.

CONCEPT

If vision is looking at the big picture, then concept is where you break it down into some bite-sized pieces. Keep the permaculture zones in mind as you consider the best place for large elements like a compost system, chickens or other animals, an orchard, a small plant nursery, greenhouse, rainwater catchment tanks, bicycle parking, sheds, a pond, comfortable seating areas, and anything else that is in your vision. Revisit the layers of the food forest to determine where best to place tall trees, fruiting shrubs, ground covers, beds of annuals and

Permaculture techniques can be used in any size garden, whether an urban lot or a suburban backyard. The key is to build healthy soil that supports a diverse community of plants.

perennials, and mushrooms. Above all, how do the systems interact and mutually benefit one another? For example, my decision to use bamboo rather than a solid wall or fence to hide the neighboring buildings has the added benefit of involving a useful multipurpose crop.

Connecting with neighbors is another way to create linkages at the concept stage. You might look for nearby gardeners from other cultures, generations, or geographic regions who are growing unfamiliar plants that you could try. Perhaps somebody is producing waste that you can recycle. A landscaper I know drops off grass clippings from across the street to help feed my compost pile. Other neighbors like to bring treats for my chickens and ducks to eat, and they bring their grandchildren to visit our little sanctuary.

ACTION PLAN

This is where you create a final plan and blueprint with a realistic budget. Think of it as a map to guide you along the right path. You already have the "why" of your permaculture garden; an action plan lays out the who, what, how, and where.

Design with as much detail as possible. Instead of writing "flowers, vegetables and fruit," indicate or list the specific varieties you want to grow and the sizes you want to plant. With specifics in place, you can begin to make your garden a reality.

Your inventory might look like this:

- 3 (4-inch) 'Munstead' lavender
- 6 (4-inch) tree collards
- 4 (4-inch) 'Egyptian Walking' onions
- 1 (3-gallon) 'Hayward' kiwi
- 1 (15-gallon) 'Hachiya' persimmon.

Also include design elements and their sizes, such as:

- 150-square-foot pond
- 500 square feet of vegetable beds
- 25 heritage breed chickens (such as Rhode Island Red, Barred Rock, or Australorp).

Plan to implement the garden in stages. First remove any existing trees and shrubs you no longer want, using the carbon harvest for fuel, construction, mushrooms, compost, or *hugelkultur*. Shape swales and drainage channels. If you have an area of lawn to convert, start sheet mulching to smother the grass and take advantage of the nitrogen it provides as it decays. If you have tough perennial weeds, see if you can borrow some goats or sheep to graze for a few weeks or sheet mulch repeatedly and dig out the unwanted weeds. Begin to improve the soil immediately, trucking in compost and woodchips if needed. Build any fences, raised beds, and trellises. Install the infrastructure for your irrigation and greywater recycling systems.

When you are ready to plant, the sequence is logical. Start with the largest layer; these are the plants that take the longest to mature. In windy sites, plant shelterbelts as windbreaks to aid the establishment of everything else downwind. In the beginning, there will be space between the larger plants as you gradually fill in the surrounding layers. Discourage weeds in the small tree and shrub layers by planting perennials from nursery containers, and lots of annual vegetables like brassicas, lettuce, radishes, and salad greens.

Just as the demographics of a particular city or neighborhood ebb and flow, the succession of plants in an ecological garden also changes over time. David Holmgren recommends starting edible forest gardens with 25 percent nitrogen fixers and eventually pruning them back to be less of a presence after the fruit and nut trees mature. Beans and peas provide food during the first season and are nitrogen-fixers, cutting back on the need to bring in fertilizer. Continue to plant and sow dynamic nutrient accumulators that bring up minerals from deep down in the soil.

RE-EVALUATION

You should re-evaluate your permaculture design annually to make sure all the separate elements weave together to make a cohesive whole. Small changes and corrections can be made, such as moving a few struggling plants to a better location or making improvements to the drainage. Thus, every year, your cycle starts again with the assessment, vision, concept, action plan and re-evaluation.

Regular re-evaluation makes permaculture designs interactive, unique and site-specific; part of an evolving process that can shift to best fit the site, your plants and human use patterns. Conventional garden designs are often static and two-dimensional. The permaculture design process is an evolutionary one that offers the opportunity to directly engage with the design, season after season.

Christopher's Garden: ACTION PLAN

In my garden, which has about 40 by 50 feet of growing space, there was a large Chinese elm that partially blocked the view of an apartment building, while still allowing sun into the garden. Underneath the tree was a small dilapidated shed. It was an ideal location for a new straw bale studio for me to dream up garden designs for clients. I planned to put a chicken coop in the back of the yard, leaving closer and sunnier areas for fruit trees and vegetables. The compost piles could go in the chicken run in a partly shady area. I chose bamboo as a fast-growing screening plant to block the adjacent apartment building and ongoing construction. It also provided trellis and fencing material, and edible shoots and forage for the poultry. The yard perimeter hosted nursery tables for my edible and native landscaping business. Herbs would be planted closer to the house for easy harvest, and there was a large area for the food forest, right on top of the lawn.

The action plan was to remove most of the ornamental shade trees, sheet mulch the lawn, build raised beds, and plant bamboo hedges and the food forest. The chicken coop was next, followed by the garden shed, the greywater system, and the rainwater collection. I've been working on this for five years and most of it is completed. There is still some work going on, with plans to remove the rest of the driveway and use the broken cement to build raised beds. The greywater system will irrigate blueberries, and more bamboo plantings can block any new neighboring buildings. Like most permaculture gardeners, I truly enjoy the process of transforming my landscape into a productive edible garden.

This straw bale studio is where I do my design work. A tall hedge of bamboo screens off the building next door. The mix of edibles in just this part of the garden includes apricots, mashua, peppers, tomatoes, tree collards, and 'Tromboncino' squash.

Christopher's Garden: REASSESSMENT

I have some culinary herbs on the deck for easy access to the kitchen, but I used to have houseplants there as well. Because my family enjoyed eating fresh fruit from the garden so much, I brought most of the houseplants indoors or gave them away. This freed up a lot of space on the deck for growing blueberries and other fruits that we frequently harvest.

In one of my garden beds, I started by planting only raspberries. The next year some sunchokes stole the show and we didn't have many raspberries to enjoy. Last year the raspberries seemed to figure out their niche along the edge of the bed. They continued to spread and yielded a good crop during the summer and fall. Now the bed contains calendula, sunchokes, a few onions, the raspberry varieties 'Indian Summer', 'September', and 'Fall Gold', a perennial 'Rocotto' pepper, and a 'Bartlett' pear tree.

It almost seems that the crops have worked out something by agreement. In June, the raspberries are fruiting. They provide a second crop in fall when the sunchokes are at their incredible heights. I harvest fewer sunchokes because I cut down the tall stalks to use as mulch around the pear tree, but I diversified my yields so now I get onions and peppers. Most of the plants are perennial, so they won't need replanting

I did have a tree collard in the bed for a while, but it never did well because it was continually shaded out by some mashua. It can be a little hard to take out something that you nurtured for so long, just like thinning seedlings, but this kind of editing helps maintain balance in the garden. Pulling out a single failing tree collard wasn't difficult because I have others in different edges of the garden that get more sun and are productive year round. Remember that in permaculture, it helps to have a fresh perspective every year. It's similar to taking a trip and then returning to realize that you see your surroundings differently.

ASSESSMENT

Mapping out the sectors on your property makes you aware of the different wild energies that affect your future garden.

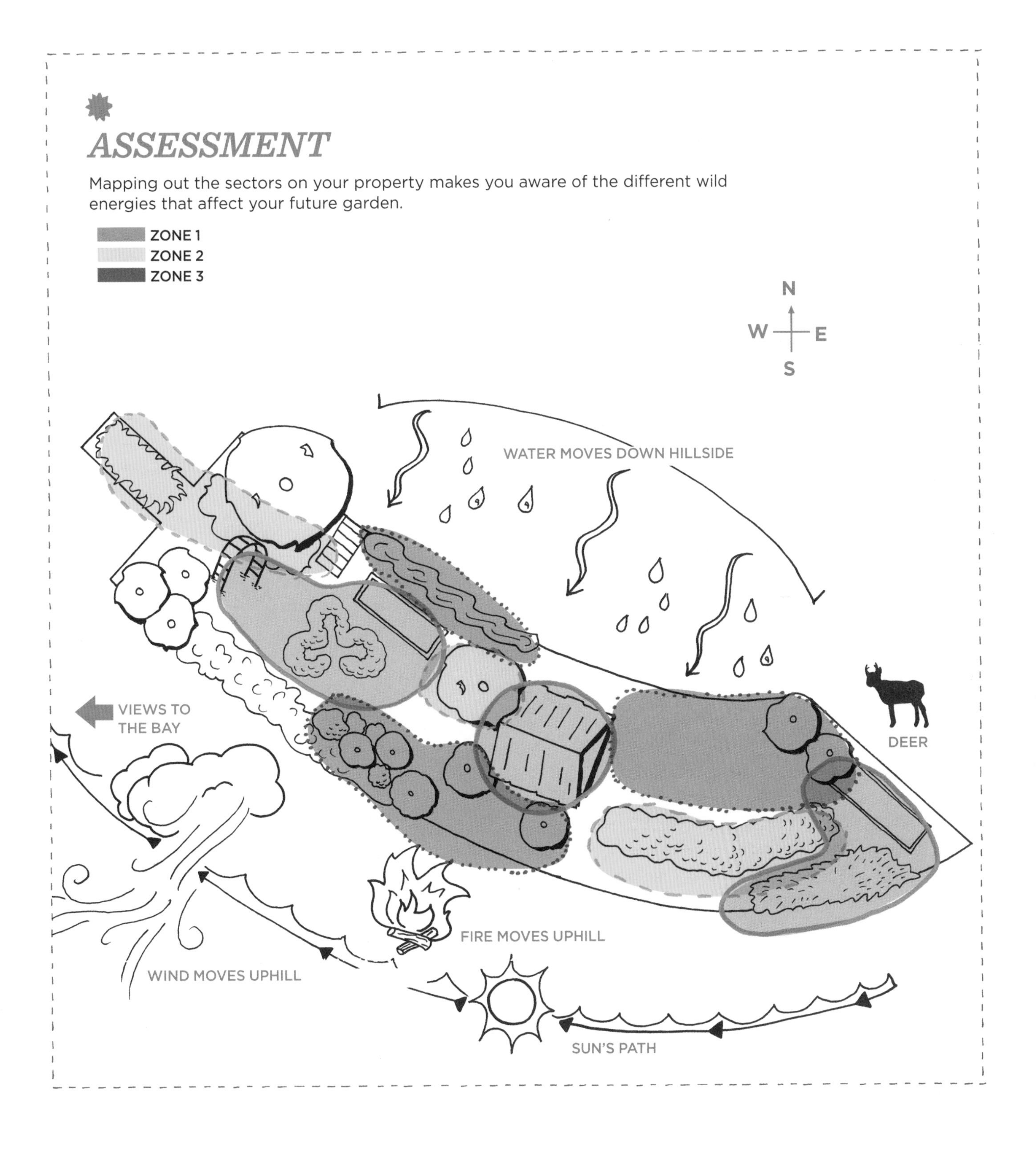

VISION PLAN

Let your imagination run wild here and come up with your dream garden. The vision is a window to your garden.

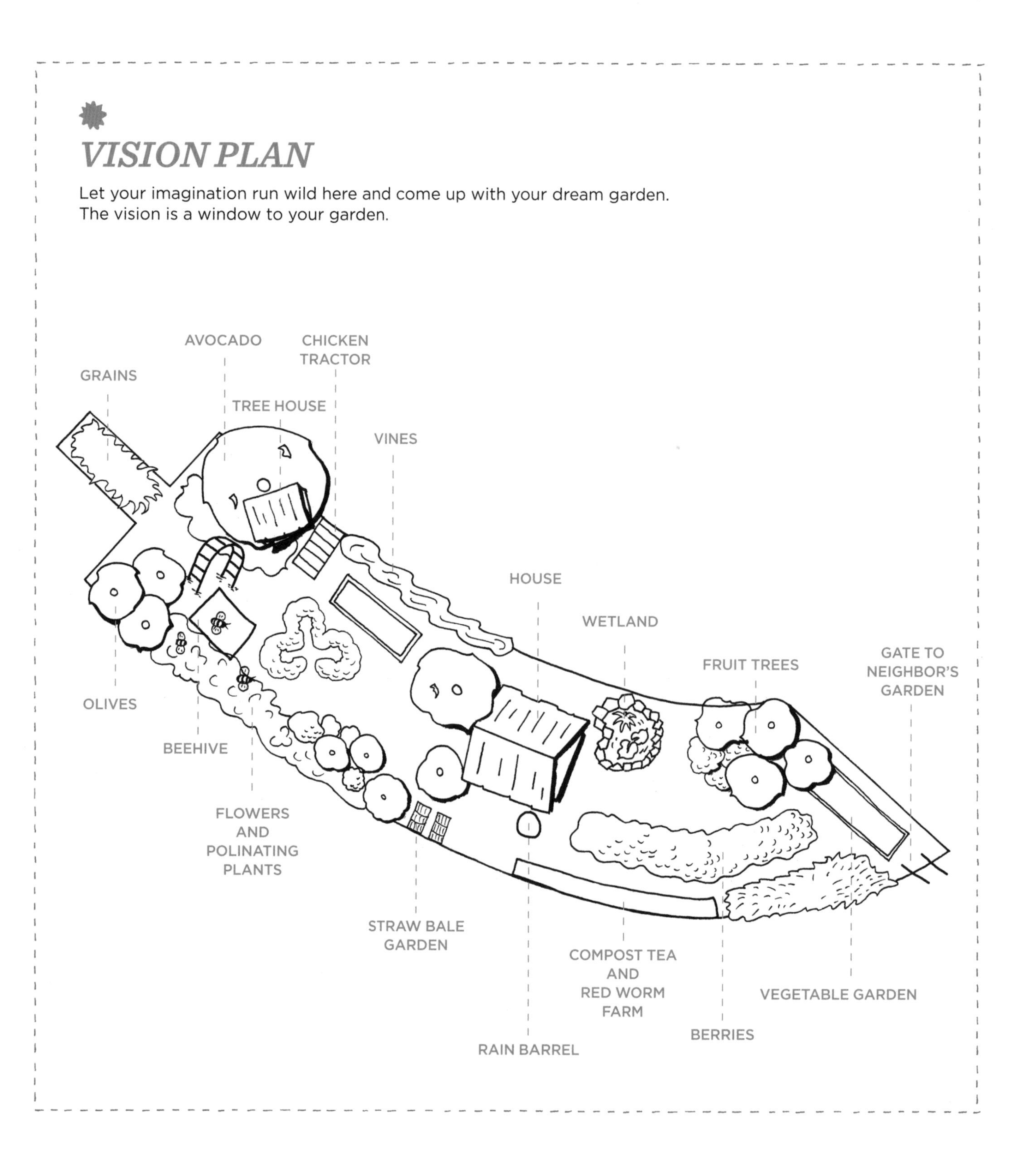

CONCEPT PLAN

In the concept phase, you can make a general map of how you plan to use the space, pruning back ideas from the vision.

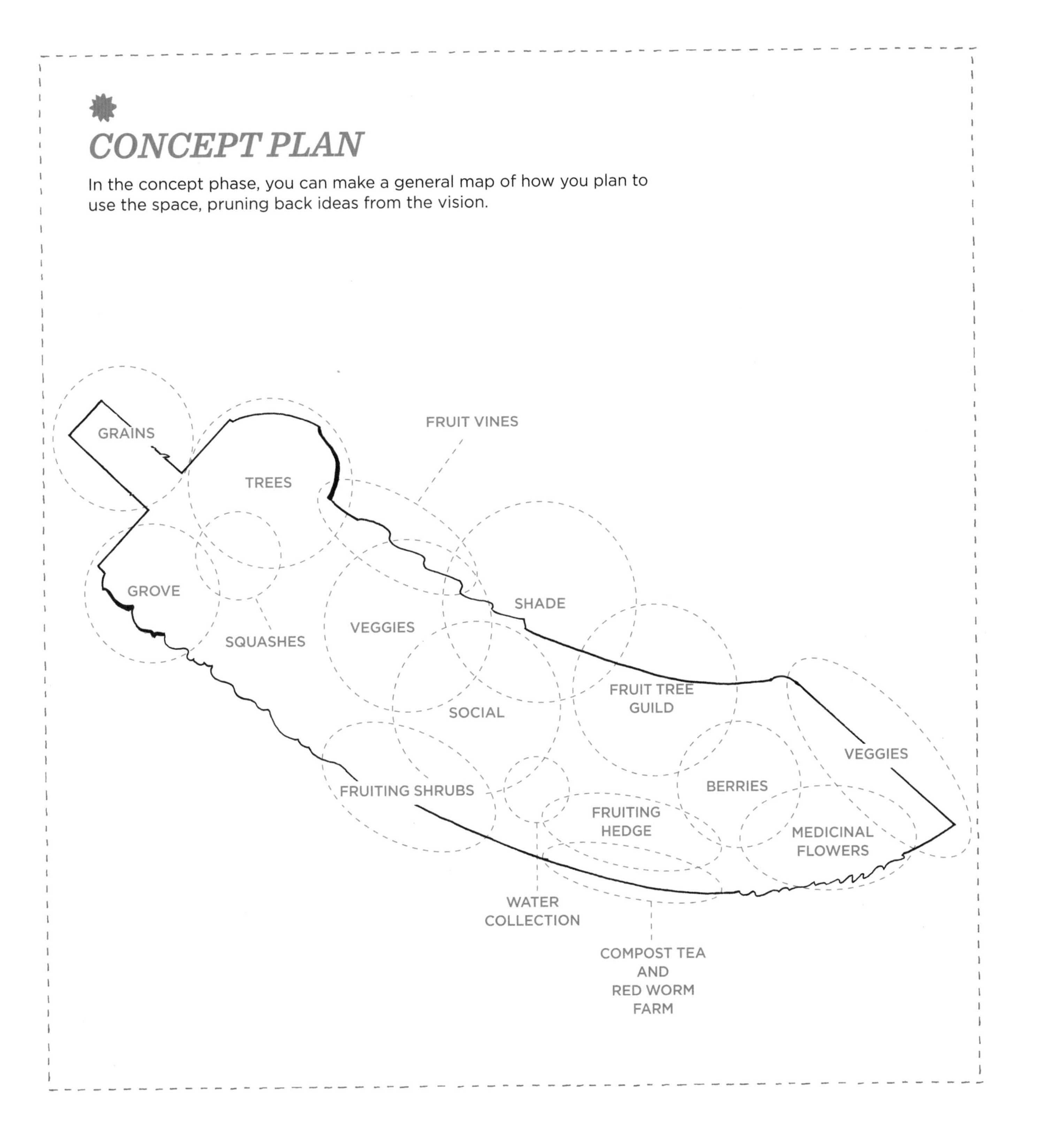

ACTION PLAN

In the action plan stage, you will add details based on your vision and concept.

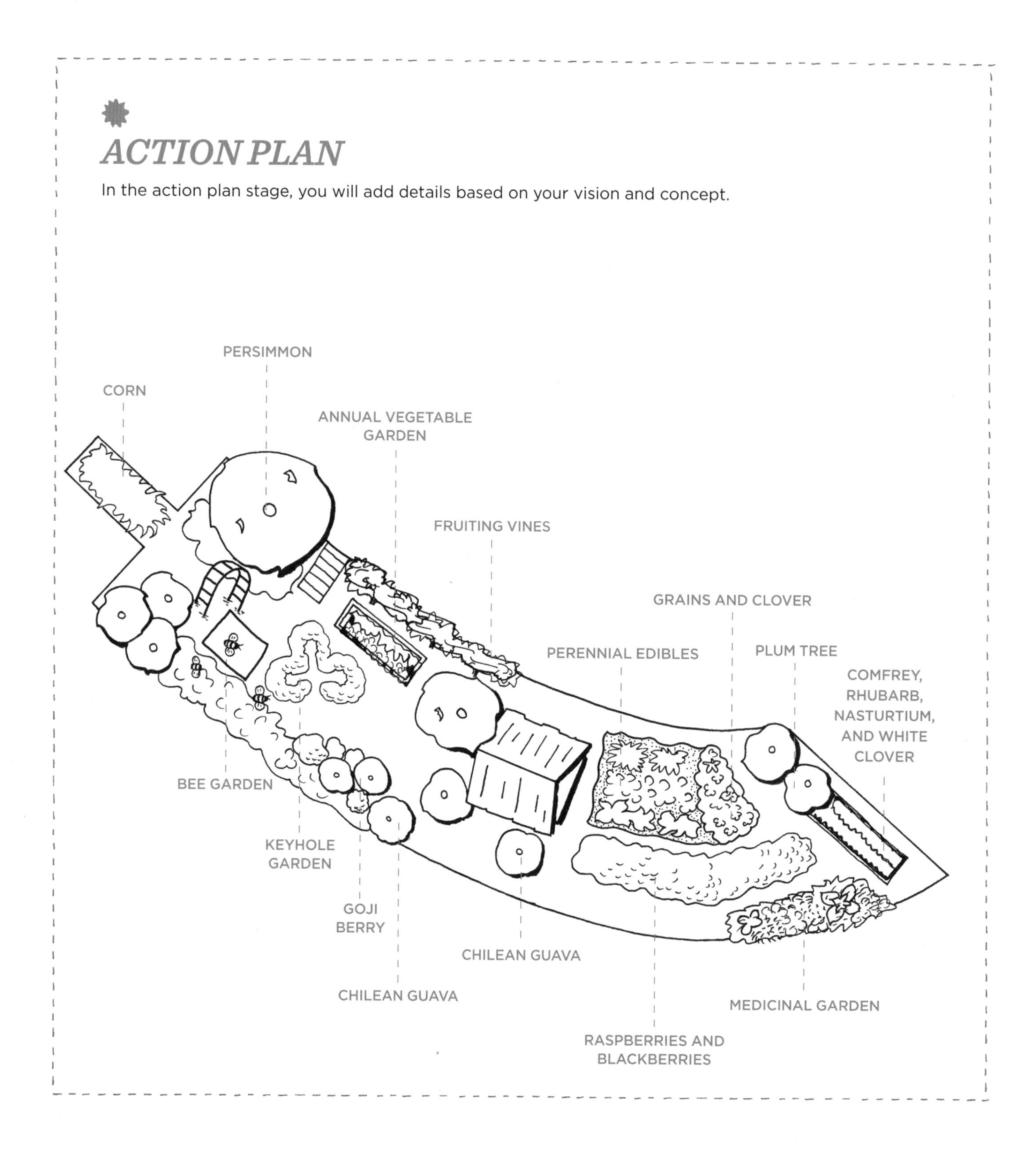

Five Permaculture Gardens

DESIGN FOR A BALCONY GARDEN

This is my 45-square-foot balcony garden in South Berkeley, California. I wanted to be able to easily access fresh kitchen herbs and cutting greens for our family of four, plus have some surplus for friends and family. The balcony is on the second story, most of it in full sun, with a trellis to block a neighboring apartment building. It's a nice spot to sit on the porch swing and watch our harvest grow.

PERMACULTURE ZONE 1

OBSERVE AND INTERACT: Many of the herbs were once planted in an herb spiral by the garden, but because of our cold nights and rainy days, we brought them up to the deck for easier harvesting.

CATCH AND STORE ENERGY: We dry and store many of the herbs for later use. The container-grown fruit nearest the wall receive extra warmth from the house.

OBTAIN A YIELD: Fruit, herbs, spices, and vegetables.

USE RENEWABLE RESOURCES: Most of the containers are recycled 15-gallon pots, but there are also salvaged ceramic pots and used wine barrels.

W N
S E

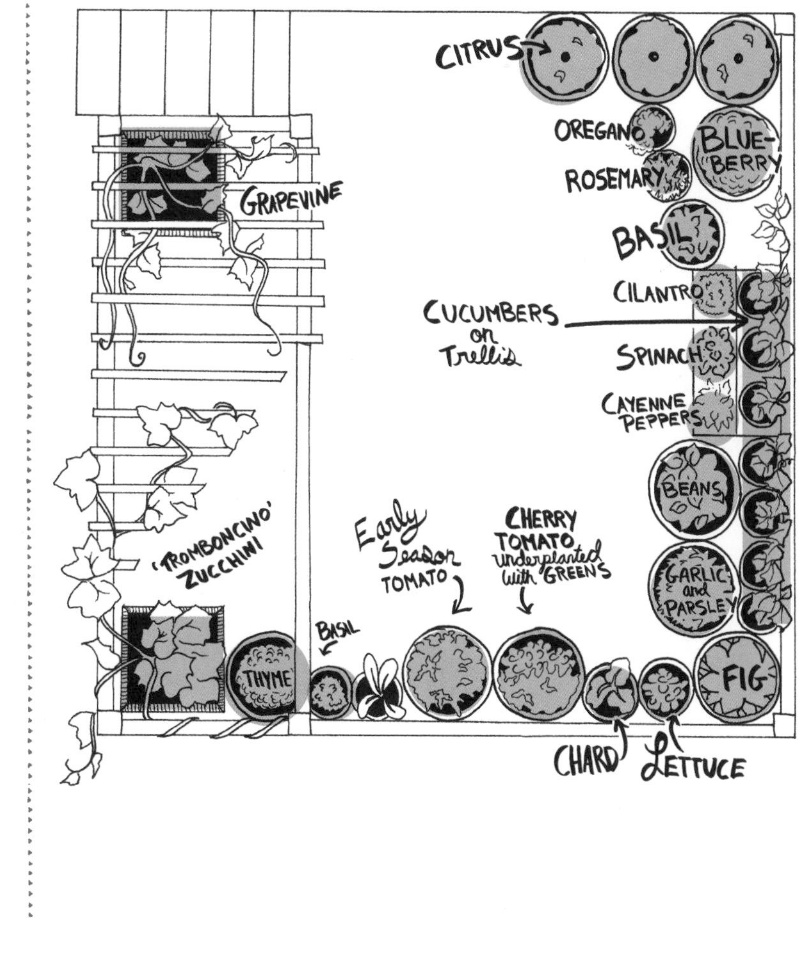

Five Permaculture Gardens

DESIGN FOR A SMALL GARDEN

This small garden belonging to Sonya and Alex maximizes space for food production and living. Two separate households share one yard.

PERMACULTURE ZONES 1 AND 2

USE THE EDGES: An adobe guest house and a wall make a space to socialize along one wall of the garden.

CATCH AND STORE ENERGY: Greywater is channeled into a constructed wetland to store nutrient-rich water. The hot tub is solar heated. The pizza oven is wood-fired.

USE AND VALUE RENEWABLE RESOURCES: Nettles, comfrey, and borage are nutrient accumulators.

OBTAIN A YIELD: Fruit trees, annual and perennial vegetables, and berries provide harvest for family and friends. Chickens and ducks provide eggs.

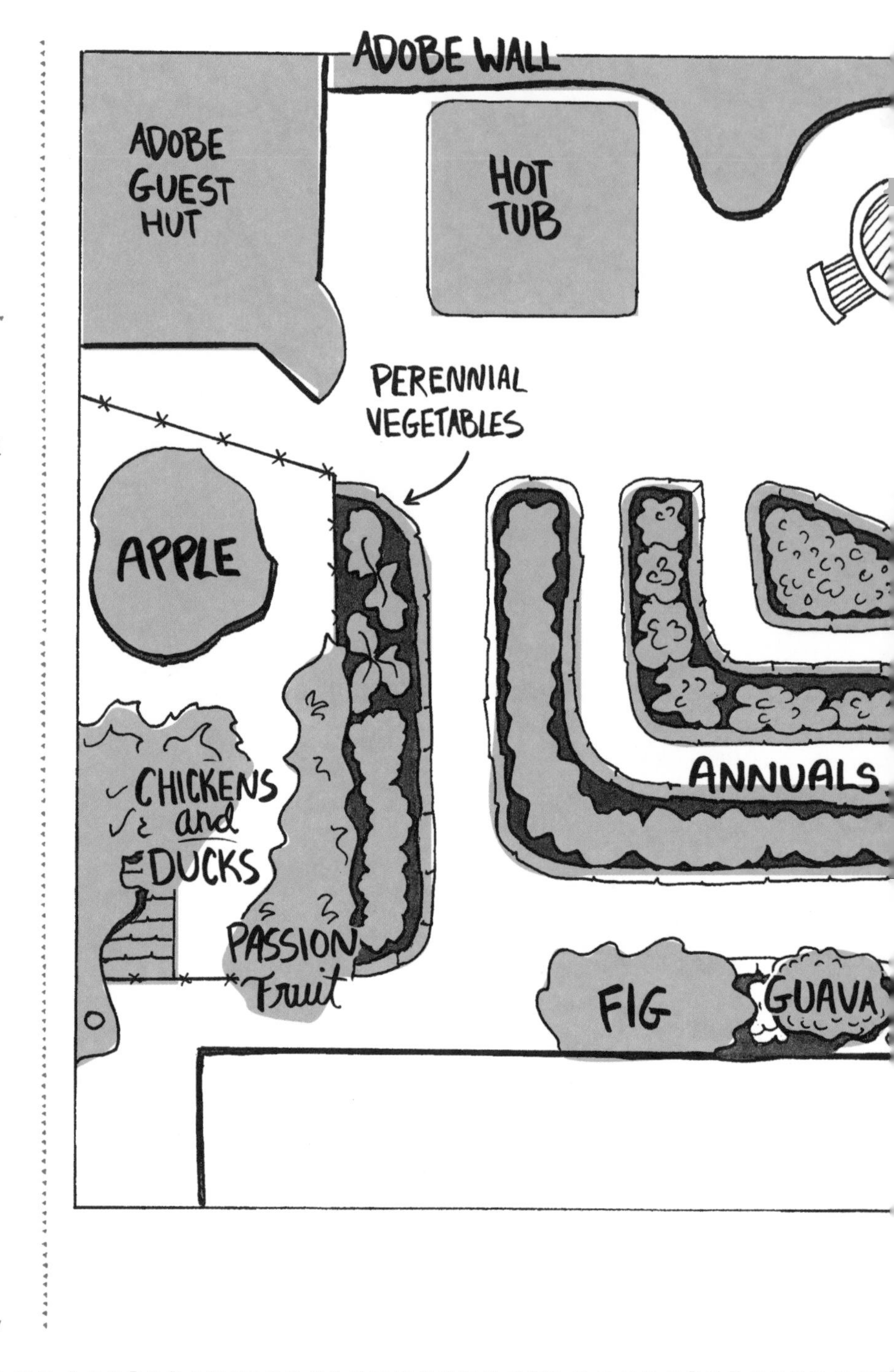

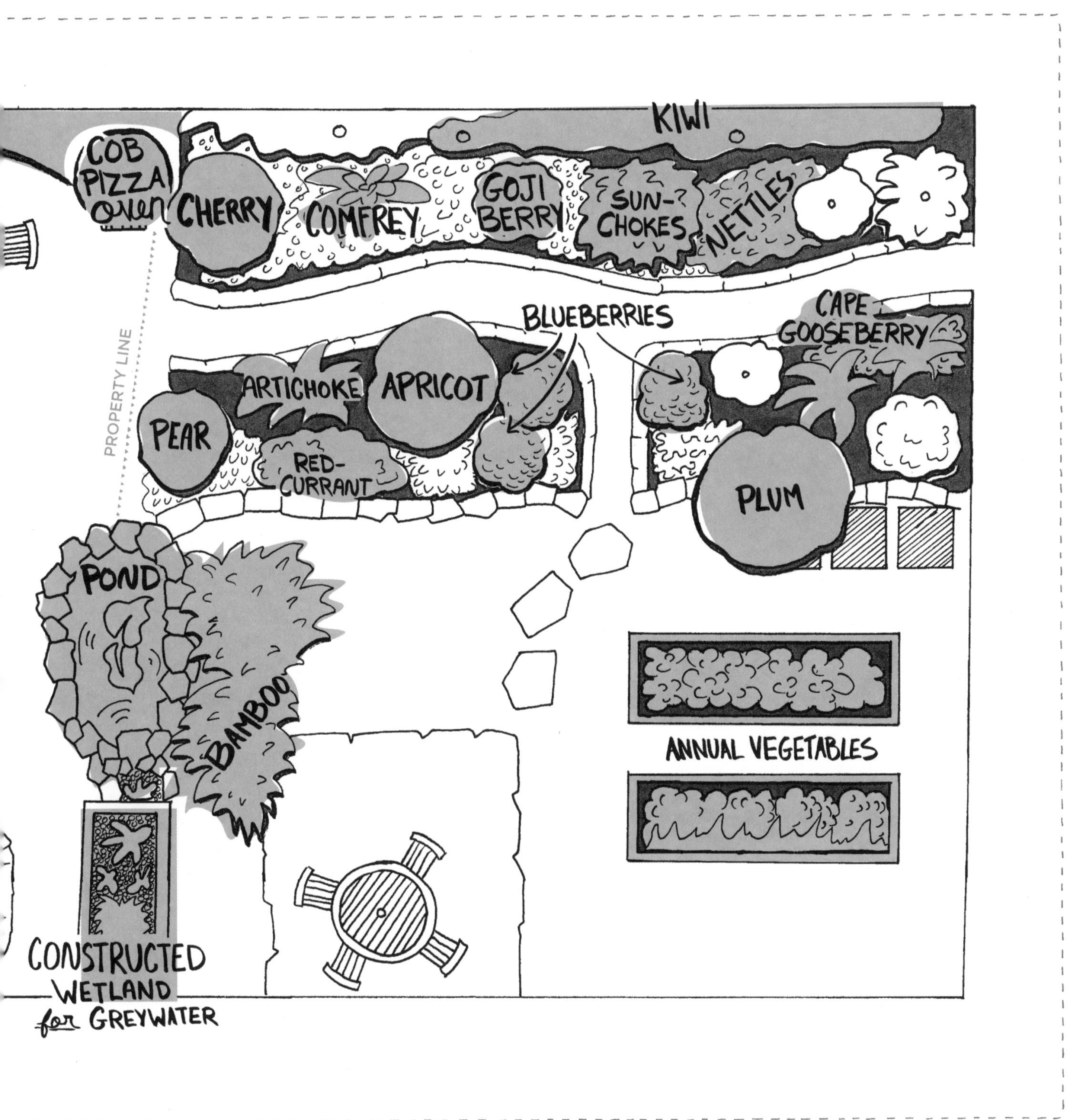
KIWI
COB PIZZA oven
CHERRY
COMFREY
GOJI BERRY
SUN-CHOKES
NETTLES
BLUEBERRIES
CAPE GOOSEBERRY
PROPERTY LINE
ARTICHOKE
APRICOT
PEAR
RED-CURRANT
PLUM
POND
BAMBOO
ANNUAL VEGETABLES
CONSTRUCTED WETLAND for GREYWATER

Five Permaculture Gardens

DESIGN FOR A WHOLE GARDEN

In this whole garden plan, you can see many of the techniques of permaculture put into practice. There are broad groupings of mixed plants, keyhole beds, and water harvesting landforms and a pond.

PERMACULTURE ZONES 1, 2, AND 3

OBSERVE AND INTERACT: The site was windy, so windbreaks of mixed plantings were used to deflect the wind.

USE THE EDGES: Trellises and windbreaks contain edible plants.

CATCH AND STORE ENERGY: Plants such as beans, grains, and pulses can be dried for later use. Legumes fix nitrogen in the soil. Trellis-grown fruit along the south-facing wall receives extra warmth in summer and drops leaves in winter to allow sun to warm the house.

OBTAIN A YIELD: Fruit, herbs, seeds, seedlings, and vegetables.

S
E W
N

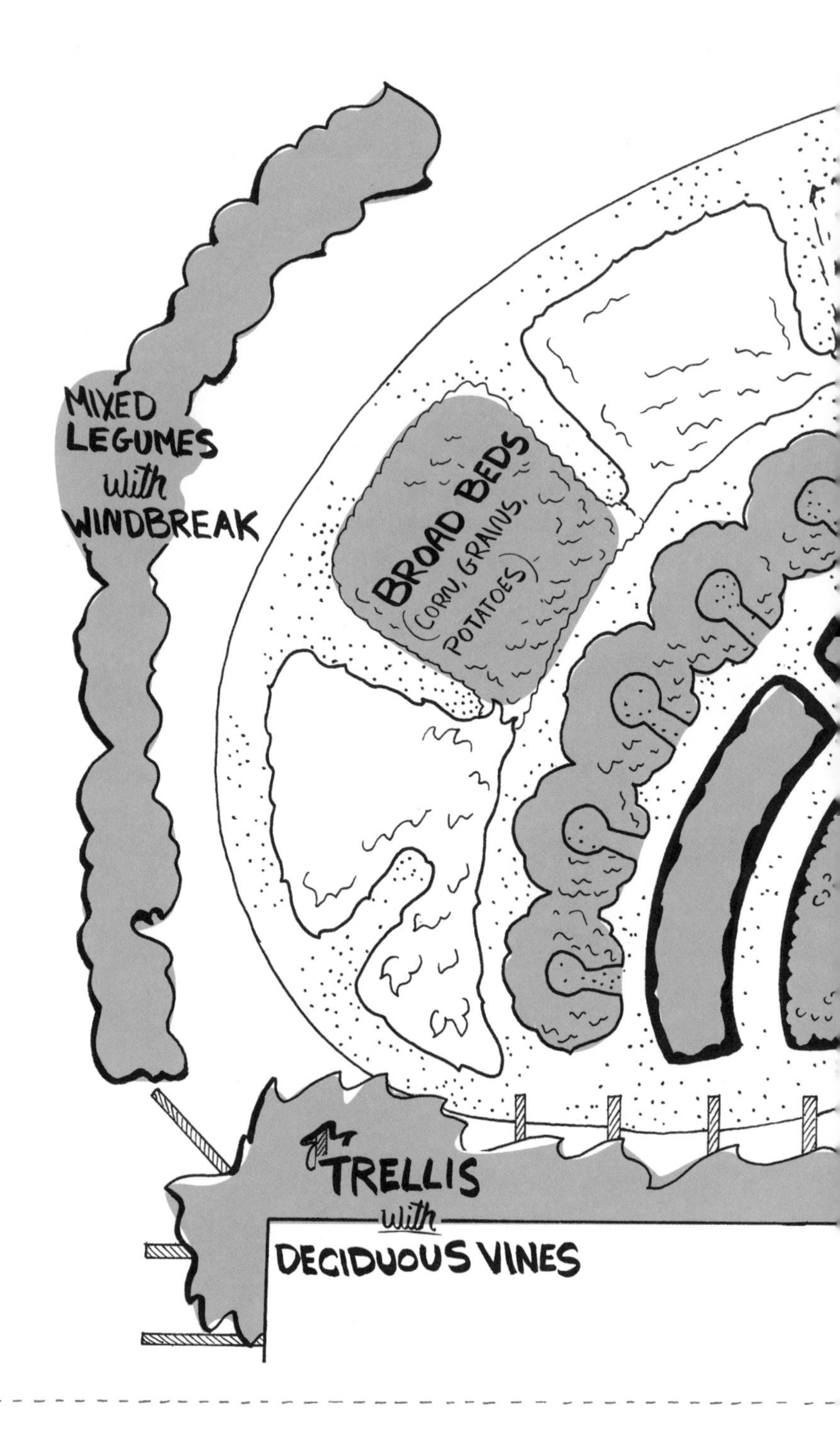

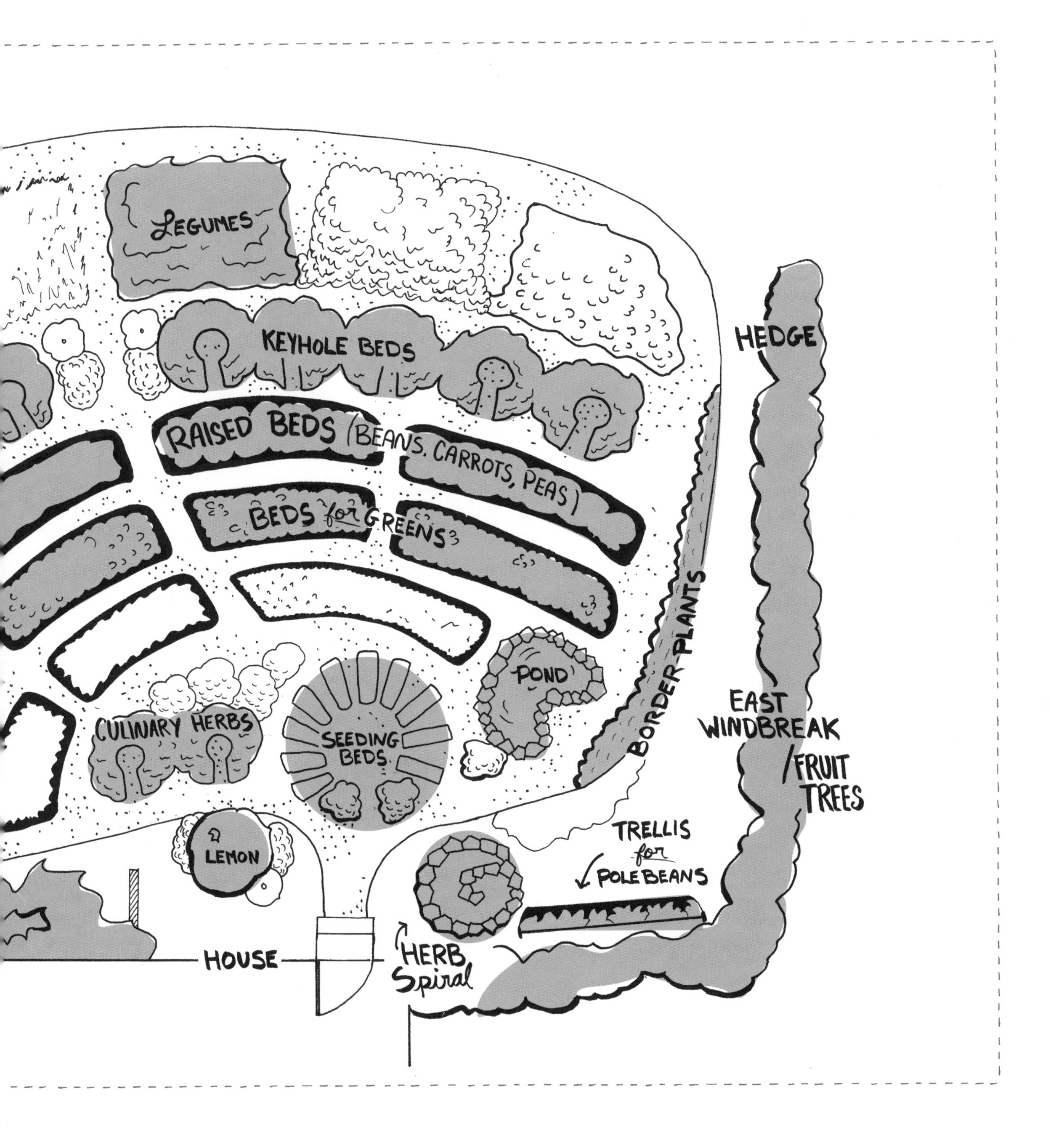

LEGUMES
KEYHOLE BEDS
HEDGE
RAISED BEDS (BEANS, CARROTS, PEAS)
BEDS for GREENS
BORDER PLANTS
POND
EAST WINDBREAK /FRUIT TREES
CULINARY HERBS
SEEDING BEDS
TRELLIS for POLE BEANS
LEMON
HOUSE
HERB Spiral

Five Permaculture Gardens

DESIGN FOR AN AVERAGE LOT

This is my garden in Berkeley. It measures about 6000 square feet, but the gardening space is less than 3000 square feet as there are structures on the property and the northeast side is very shady. We wanted to have fresh chicken and duck eggs, space to socialize and work with friends and family, and a kitchen garden with fruit, flowers, herbs, and vegetables. A nursery space was also important.

PERMACULTURE ZONES 1 AND 2

USE THE EDGES: Bamboo hedges block the large buildings next door. Pollinating, drought-tolerant natives are planted on the shady northeast side. Keyhole beds, mandala beds, and herb spirals maximize planting space.

CATCH AND STORE ENERGY: Rainwater harvesting to catch water. Lots of mulch to act as a sponge for rainwater (22 inches per year).

USE AND VALUE RENEWABLE RESOURCES: Chickens and ducks help control slugs and snails and give use manure. We save seeds and trade them with BASIL members. Bamboo hedges make construction timber.

OBTAIN A YIELD: Fruit trees and berries provide harvest for family and friends. Chickens and ducks provide eggs.

PRODUCE NO WASTE: Our household waste is converted to soil. I do a weekly pickup of vegetable scraps from a nearby restaurant to feed the fowl, and make into compost.

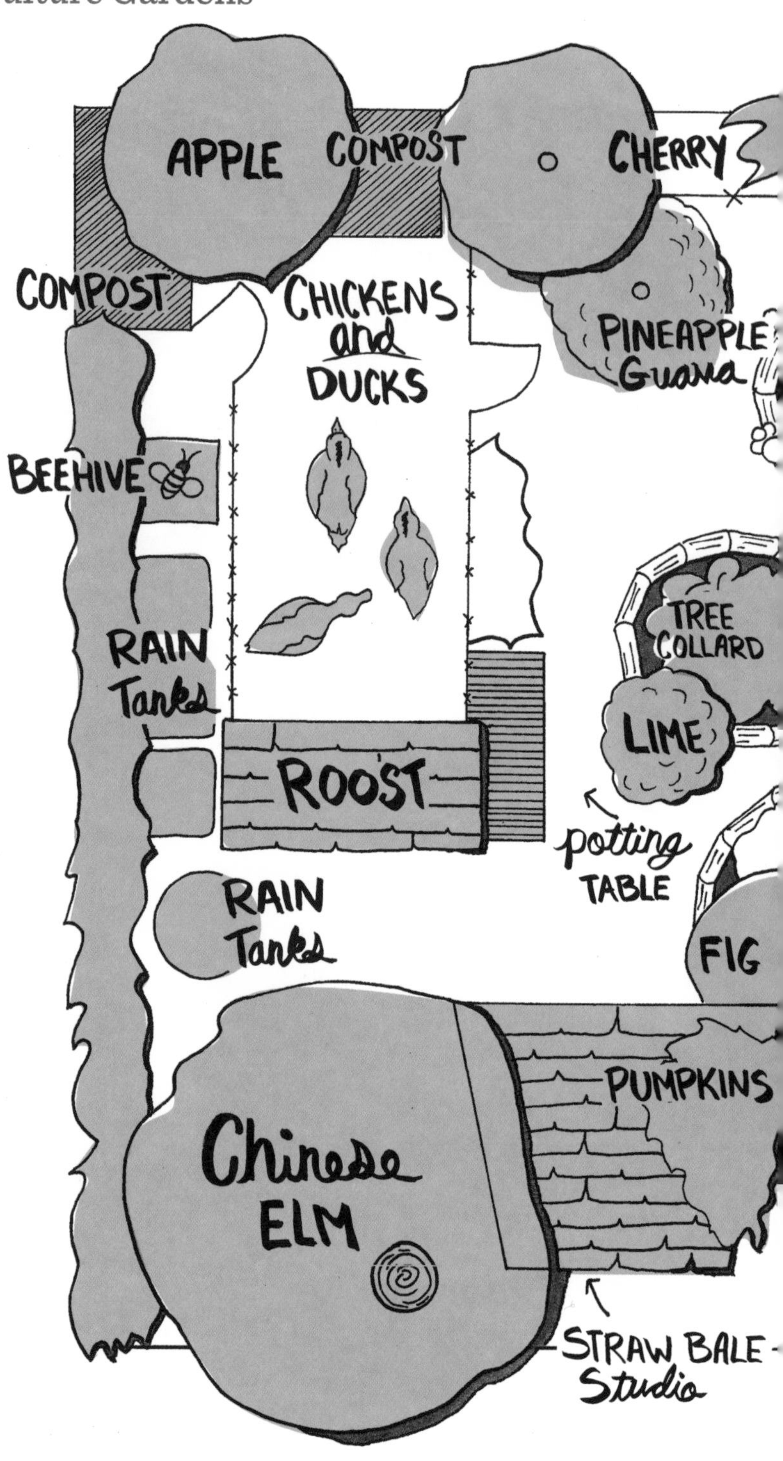

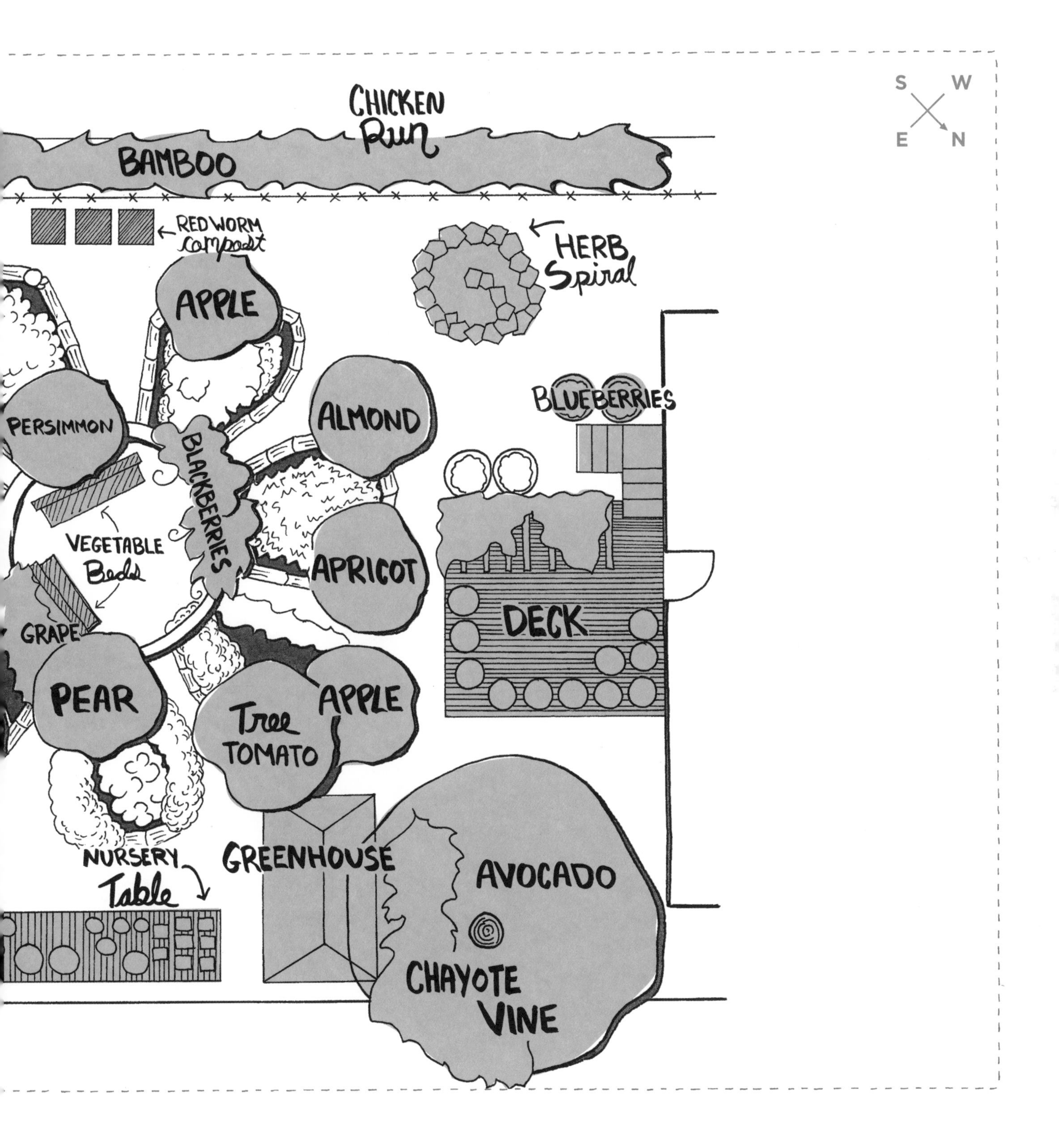

S
W
E
N
CHICKEN Run
BAMBOO
RED WORM compost
HERB Spiral
APPLE
PERSIMMON
BLACKBERRIES
ALMOND
BLUEBERRIES
VEGETABLE Beds
APRICOT
DECK
GRAPE
PEAR
Tree TOMATO
APPLE
NURSERY Table
GREENHOUSE
AVOCADO
CHAYOTE VINE

Five Permaculture Gardens

DESIGN FOR LARGE URBAN LOT

This lot in East Oakland is a half-acre (21,500 square feet). It's in a warm banana belt, so many heat-loving crops can be grown. The exposure is mainly western, with a view of San Francisco Bay. The neighborhood has many homes with large lots. The owner is a nurse who lives on the site with her two sons. She wanted a healing sanctuary, a place to raise animals and grow plenty of vegetables and fruit. It's a true mini farm cared for by the family and a land partner.

PERMACULTURE ZONES 1, 2, AND 3

OBSERVE AND INTERACT: The seasonal creek was used to catch and store water.

USE THE EDGES: Windbreak legumes and hedges provide privacy. Blueberries and raspberries form the longest boundary of the garden. Keyhole and mandala beds maximize space.

CATCH AND STORE ENERGY: Legumes fix nitrogen in the soil. Chickens, ducks, and rabbits provide onsite manure and pest control. Swales and a pond harvest water.

OBTAIN A YIELD: Fruit trees provide harvest for family and friends. Chickens and ducks provide food. Goats provide milk. Many annual and perennial crops.

INTEGRATE RATHER THAN SEGREGATE: The labor is shared by family living nearby and a land partner.

USE SMALL AND SLOW SOLUTIONS: The buildings and growing beds started small and manageable. These are beginner gardeners building on their successes.

USE AND VALUE DIVERSITY: The families are multicultural, reflecting the neighborhood and city in mutually beneficial ways. Residents actively seek out elders in the neighborhood for experience and wisdom.

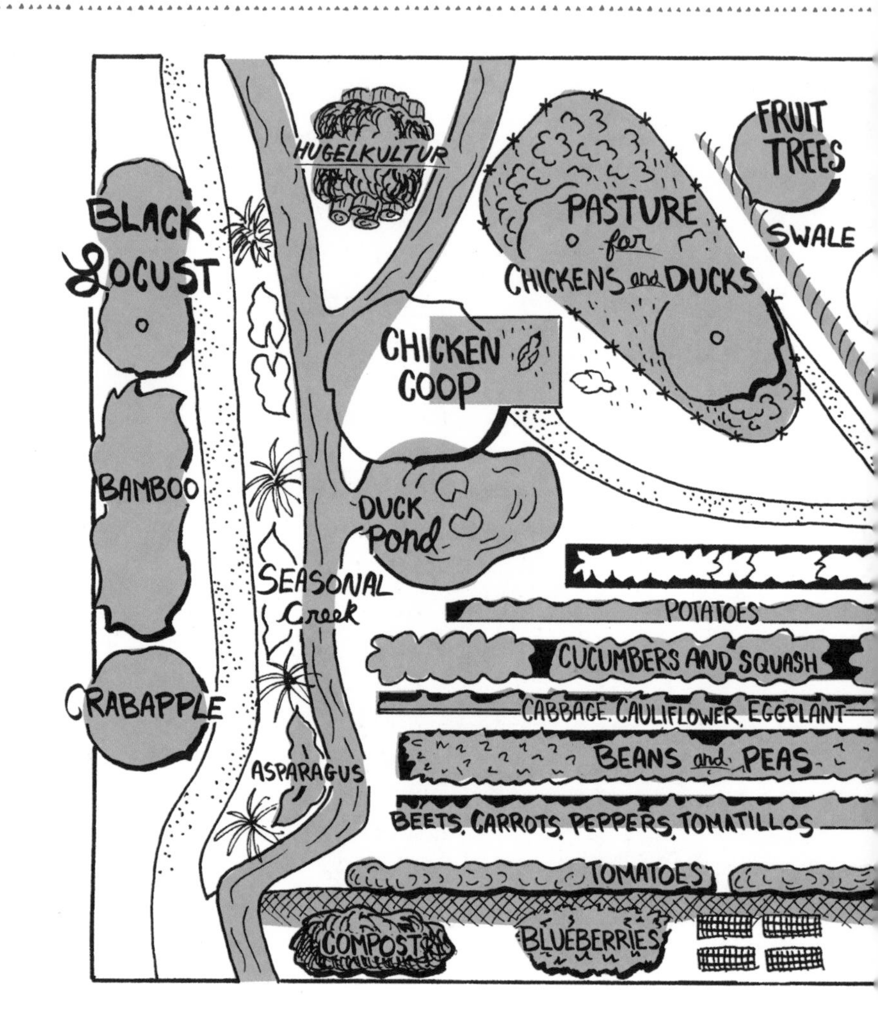

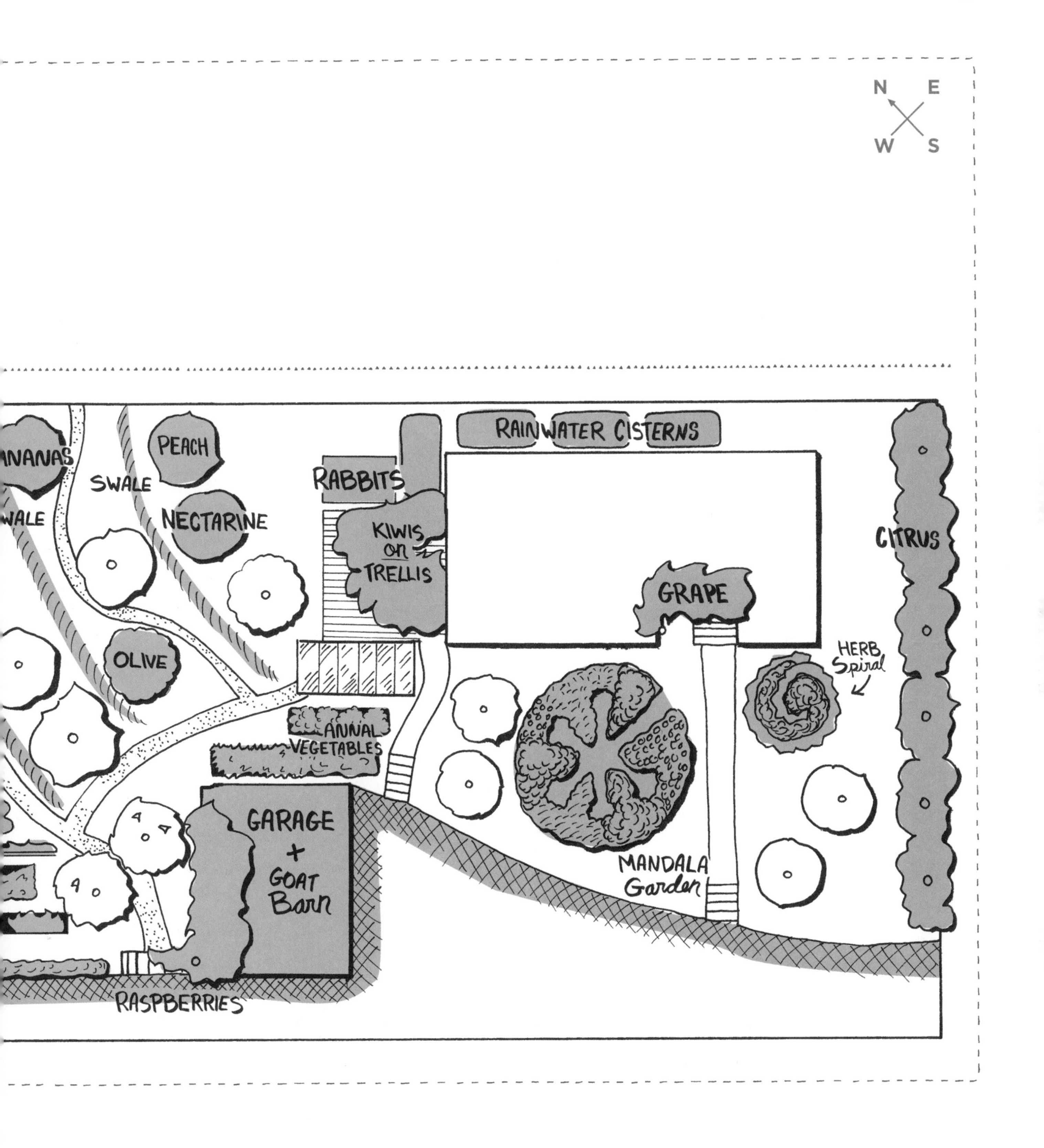

N
E
W
S
RAINWATER CISTERNS
PEACH
SWALE
RABBITS
NECTARINE
KIWIS on TRELLIS
CITRUS
GRAPE
OLIVE
HERB Spiral
ANNUAL VEGETABLES
GARAGE + GOAT Barn
MANDALA Garden
RASPBERRIES

Compost completes the cycle of growth and regrowth in the garden, from soil to growing plant and eventually back once more to soil.

4 BUILDING THE SOIL

Soil Basics

THE SAYING "YOU ARE what you eat" is true for people, and it is equally true for soil. Soil can be healthy only if it is provided with nutrition. The condition of the soil is grounded in the ethical principle of earth care and the love of all things related to soil, soil building, and regeneration.

Soil texture results from millions of years of the breakdown of rocks and plant and animal matter. Depending on the predominant particles that remain, soil tends to be sandy, clay, or loam, or a mix of the three. Soil structure is different. It refers to the way soil holds together, and depends on the amount of air, moisture, and organic matter it contains. If you try to dig soil that is saturated with water, it becomes compacted. If soil has too little water, it turns to powder. When soil has a good structure, it falls apart into pea-sized pieces when you dig, making it easy to work with.

In a natural forest, it can take up to one thousand years to make an inch of topsoil. If there are no plants and animals to give back to the soil through decomposition, the ground will eventually turn to lifeless dust. Plants are the foundation for the dynamic community of organisms in the soil food web, converting sunlight into green biomass, which is then consumed by other living things including bacteria, fungi, protozoa, nematodes, insects, reptiles, birds, and animals. Maintaining diverse and healthy soil is the best practice for raising healthy plants.

Modern agriculture has taken the natural cycling of matter and energy out of the equation, which has led to the loss of topsoil at an alarming rate. According to Kenny Ausebel, the founder of Bioneers, 75 percent of the continent's topsoil has been lost

Soil is the basis of all growth, and edible plants demand good soil. In permaculture, we continually enrich the soil by continually adding organic matter back in to be recycled by soil organisms.

since the introduction of European-style plow agriculture.

We cannot strip-mine soil and expect it to produce healthy food. To keep soil alive and healthy, and for our gardens to be truly sustainable, organic material from plants and animals must continually be reintroduced. As gardeners, we have a huge impact on the quality of soil health and can greatly impoverish or enrich the soil by our gardening and farming techniques. Composting, mulching, using cover crops, rotating crops, and adding compost tea all serve to enrich the soil food web because they mimic the natural processes. These methods of building and maintaining healthy soil are easily learned, and result in healthy soil to feed our edible gardens.

In community gardens where there is plenty of space, permanent compost beds can be laid out in long rows. These windrows convert brown and green organic matter into an ongoing source of compost.

Compost

Composting exemplifies so many of the basic permaculture principles—work with nature, produce no waste, use and value resources, catch and store energy—that it is impossible to imagine sustainable gardening without it. The process is based on the natural cycle of decay and regrowth, where plants at each stage of their life cycle contribute to the soil food web by dropping leaves, fruit, stalks, or flowers to be broken down by soil-dwelling insects and microorganisms, and finally to provide nutrition for the next generation.

COMPOST BASICS

Four elements are needed to make a compost pile: green (wet) material (including kitchen scraps), brown (dry) material, air, and water. In general you want one third kitchen scraps, one third green material, and one third brown material, making sure there is enough air for circulation and the material is not too moist nor too dried out (the whole pile should be as damp as a wrung-out sponge). Too much green material or water make compost anaerobic and putrid. Too much air dries it out so that no biological decomposition takes place. If too many browns are added, the pile won't heat up. Just as with any recipe, it takes practice to correctly balance the ingredients.

The bin size depends on the volume of compost you are processing, and the amount of time and energy you have. The easiest type of bin to make is a circular wire structure; it also requires the least turning and can be harvested a few times a year. This is a good option for processing a household's daily kitchen scraps. Stackable plastic bins are convenient and you can turn the compost frequently for a faster yield. The three-bin system—three 3-cubic-yard wooden bins in a row—is ideal if you have the space and the enthusiasm for turning piles.

Another type of compost system is windrows, where materials are stacked and laid down in long rows, like an ever-growing snake. They are not turned, and are harvested once or twice a year. Windrows are large scale and are suitable for the home gardener only if you have lots of room and a reliable source of green and brown materials.

For dry or hot climates, compost needs to be well shaded, and it needs good covering for moisture protection. You can protect the pile with mulch and tarps. For cold climates, you don't want to trudge through too much snow to get to the pile, so you might put it in zone 1 or 2. Covering piles with straw bales can keep them from freezing. A removable wooden lid on an open pile can prevent large snowdrifts from accumulating on top. In wet climates, make sure there is protection from the rain. A soggy wet pile will have no air and will become anaerobic, producing an unpleasant smell.

BUILDING THE COMPOST

Some piles will have more office paper and home-generated vegetable scraps; others will have more leaves, animal manure, and garden clippings.

The more variety you put into your compost, the better the end product. Look outside your own garden for sources of material to supplement your pile. Places like grocery stores, juice bars, coffee shops, restaurants, and food banks often throw out spoiled produce and may welcome your offer to take the waste off their hands. Farms or stables may have excess manure.

Ask your neighbors if they compost and if not, tell them you'll gladly accept their kitchen scraps—but be consistent and pick up their compost on a regular basis. You are demonstrating the permaculture strategy of turning waste into a food solution. If you have an abundant harvest, share it with those who have contributed.

Layer one part browns, one part greens from the kitchen, one part greens from the garden, and repeat until your pile is 3 feet tall. In between the layers, water the pile to the dampness of a wrung-out sponge. Keep adding 4-inch layers of greens and 2-inch layers of browns every few weeks. Turning the pile with a garden fork speeds decomposition. An efficient compost pile generates heat as the material decomposes. The ideal temperature range is 130° to 150° F. Maintaining the pile at this temperature for about a week will kill many weed seeds and pathogens.

You know your compost has finished decomposing by the look and the smell. There won't be any recognizable ingredients like vegetable scraps and straw, and it will have a dark, earthy color and smell rich and sweet.

CARBON-TO-NITROGEN RATIO

Bacteria build their bodies with carbon and use nitrogen as their fuel. This bacteria is what will help to break down your kitchen scraps and straw into compost, so you want to promote the right conditions for their growth, which includes a ratio of carbon to nitrogen (C:N) of about 30 to 1.

Gitanjali takes shredded office paper to add to the compost.

BROWN COMPOST MATERIALS

- cardboard
- dried crop residues like sunflower stalks or chaff from seed winnowing
- dried leaves
- newspaper
- office paper
- sawdust (avoid plywood and treated wood)
- spoiled hay
- straw
- wood shavings
- woodchips

GREEN COMPOST MATERIALS

- animal bedding (chicken, cow, goat, horse, rabbit, and sheep)
- coffee grounds
- compost crops like comfrey
- cover crops
- crop residue (such as spent tomato vines)
- fish bones
- garden weeds (without mature seeds or perennial roots)
- grass clippings
- juice pulp
- kitchen scraps
- manure (chicken, cow, goat, horse, rabbit, and sheep)
- restaurant trimmings
- seaweed
- tea bags and leaves

CARBON-TO-NITROGEN (C:N) RATIO

GREENS	BROWNS
juice pulp **35:1**	sawdust **500:1**
garden weeds **30:1**	cardboard **350:1**
fresh leaves **30:1**	newspaper **175:1**
horse manure **25:1**	woodchips **100–500:1**
kitchen scraps **25:1**	straw **75–125:1**
restaurant trimmings **25:1**	dried crop residues like corn stalks **60–75:1**
used coffee grounds **20:1**	dry leaves **50:1**
seaweed **19:1**	hay **25:1**
cow manure **18:1**	
fresh grass clippings **12:1**	
chicken manure **7:1**	
fish scraps **5:1**	

These numbers are a guide to the C:N ratios of various compost materials. I don't have my calculator with me when I'm building a pile, but if I see that dried leaves are 50:1 C:N, and sawdust is 500:1 C:N, I know to use sawdust sparingly. Horse manure and kitchen scraps are both at 25:1, so I use them liberally. On the other hand, I use grass clippings (12:1) and chicken manure (7:1) sparingly in thin layers.

RED WORM COMPOSTING

Converting organic matter into finished compost through a red worm farm is called vermiculture. *Eisenia fetida,* or red worms (also called manure worms, composting worms, red wigglers, tiger worms, or gourmet worms), are primary decomposers that live on the surface of the soil and eat leaf litter and manure. Vermiculture produces no waste and is ideal for people with little composting space or who find the idea of turning compost too challenging.

The finished product—worm castings—is an incredibly rich plant amendment. It contains beneficial enzymes and microbes that increase fertility and nutrient uptake in your plants. The castings actually pass through the red worms' digestive systems several times. Claudia Taurean, known as the Worm Lady of the East Bay, says, "[Red worms] eat what they live on and live in what they eat." The worms multiply so prolifically that eight adult worms can breed into 1500 in six months. You can give the extra worms to friends and family, sell them, or feed them to the chickens and ducks.

Red worm bins can be made out of many containers, even a salvaged food-grade tub will work. Punch or drill air holes on the sides near the top (but not on the lid). Place some bedding, such as shredded paper, in the bin. Red worms can eat half their own weight daily; 1 pound of red worms will eat at least 3 ½ pounds of material a week. They will eat all of the bedding material, as well as

Kitchen scraps add green, wet matter to the compost pile. Be sure to balance these greens with dry brown materials or the compost will not decompose properly.

LEFT
Vermicomposting is not done by earthworms, but by red worms. The result is a rich compost and a liquid that can be brewed into a nutrient-rich tea.

RIGHT
At Oak House Permaculture in Shropshire, UK, worm juice from a wormery's processing of vegetable waste gets watered down with rainwater in a watering can for a simple liquid fertilizer to use on veggies.

kitchen scraps, crushed eggshells, and coffee grounds. Do not add meat, dairy, grease, sugar, citrus, or onion peels. To avoid the material becoming too acidic, a good ratio of kitchen scraps is 2:1 vegetables to fruit. To prevent flies from breeding, cover the kitchen scraps with 6 inches of dry material like newspaper or dried leaves. You must harvest the castings regularly because they do become toxic to the worms in time.

It's important where you put the bin. Don't locate it in the full sun or the worms will dry out and die. And they can freeze, so keep the bin in a sheltered place during the winter. The ideal temperature is 55° to 77° F. In colder climates, protect the bin from frost by placing it under a large evergreen tree, shrub, or bamboo. I keep my worm bins in a sunny area with easy access to a path, and close to the larger composting system. It is easy to harvest the castings with the bin right in the garden.

After two months, you can harvest the worm castings and the liquid (worm juice) that collects at the bottom of the bin. The simplest way is to flush water through the bin, and strain out the live red worms from the liquid. Reintroduce the red worms to the bin with fresh bedding and regular food.

The liquid can be used to make a nutrient tea for your garden plants. Dilute the worm juice until it turns the color of tea (not coffee), typically about ten parts water to one part juice. Use the resulting tea immediately, pouring it from a watering can around the base of your plants. For an even more nutrient-rich solution, make compost tea. If you use freshly harvested worm castings as a soil fertilizer in containers or in garden beds, mix the castings with compost so they make up only about 10 percent of the total volume. Castings are an excellent topdressing, but they will dry into a hard crust if you don't mix them with compost or cover with mulch.

Protect outdoor bins from raccoons, opossums, rats, and snakes, as they all love to feast on worms and scraps. An easy way to do this is to keep a brick on top of your worm bin, or cover the exterior with hardware cloth or chicken wire.

Mulch

To learn how to improve the quality of our soil, we can observe how the forest ecosystem does it naturally. When leaves, twigs, and branches fall there are no landscape maintenance companies to dispose of them. Instead, they serve as a blanket to suppress weeds and to keep the soil from eroding or drying out. Ultimately, with the help of worms and soil organisms, they decay to become nourishment for future growth. In the garden, we mimic this onsite recycling by mulching.

In my garden in northern California, there are only 22 inches of annual rainfall, most of it falling from November through May. During the dry summers, if the ground is left exposed, the sun bakes the clay soil into adobe, and digging is like working with concrete. Adding a layer of mulch—minimally 4 inches per year—allows the soil to remain moist and workable. You can see the life in the soil food web moving and working in the soil. Because mulch keeps down weeds, maintains soil moisture, and builds fertility as the material slowly decomposes in place, it's a labor-saving tool. Mulching with woodchips is one of my favorite garden techniques, because I really don't like to weed. The only drawback to mulch is that it can harbor scavenging insects, and slugs and snails. In spring, keep the mulch away from tender new seedlings, but in winter encourage your ducks and chickens to peck through it.

Good mulching materials include bark, cardboard, finished compost, grass clippings,

Ruth Stout, Gardening Rebel

In the 1930s, Ruth Stout and her husband left the city for a rural life in Connecticut. Back then, there was only one plowman with a tractor in her area, and she was impatient to start planting her one-acre food garden. So she asked her asparagus plant, a perennial that needs no tilling, why she needed to plow the rest of the garden. She recounts that it told her, "You don't need to."

This began Stout's life as an organic gardening pioneer. For the next fifty years, she went against conventional wisdom and never again dug her garden beds. Instead, she practiced low-work and low-input gardening. She developed a mulching system in which she used a thick layer of straw to keep down weeds and slowly build up fertility as the straw rotted. Like Fukuoka, she hardly ever used compost, relying on her onsite plants and the straw mulch. She was a writer for *Organic Gardening* magazine and the author of *How to Have a Green Thumb Without an Aching Back: A New Method of Mulch Gardening* and *The Ruth Stout No-Work Garden Book: Secrets of the Famous Year-Round Mulch Method.* While not enough credit is given to Ruth Stout for her contribution to permaculture, she is widely known in organic gardening circles for these mulching methods.

Making Compost Tea

Compost tea is made of red worm tea, molasses, and water. The blend is left to culture with an aquarium bubbler for three days and is then ready to apply to your plants as a foliar spray, rich with vitamins and minerals as well as beneficial bacteria and yeasts. The tea helps to prevent certain diseases and can help plants bounce back after pest damage.

Materials

5-gallon bucket
1 pound of red worm castings or 4 cups of red worm tea
6 cups of molasses
water
aquarium bubbler
fabric paint strainer
garden pump sprayer

Instructions

1. Place red worm castings or tea in the bucket and add 3 gallons of water.
2. Add molasses, stir.
3. Insert aquarium bubbler, leave for three days.
4. Strain through a fabric paint strainer and pour into a pump sprayer. Apply immediately to plants. Compost tea is good for only one day.

Choose a wind-free day to apply the tea. Try to cover both sides of the plant leaves. To prevent any spray from entering your nose or mouth, wear goggles and a mask.

leaves, newspaper, straw, spoiled hay (don't use this as a top layer because it will sprout), tree trimmings, well-rotted manure, and woodchips (not pressure treated and not from plywood).

Where can you get mulch materials? Everywhere! From your own yard, to your neighbors' yards, in bags on recycling day, or from local sources like farmers, tree trimmers, woodshops, lumberyards, mills, and soil and mulch companies. Your municipal works department may offer mulch or leaf mold. Some municipalities maintain piles of these materials, so that residents can load up their cars and trucks. The city saves time and money by not having to process the tree trimmings and fallen leaves in their facilities. The City of Berkeley has been doing this for years and now many other cities have jumped on the bandwagon. You can also look on the internet—websites like Craigslist and FreeCycle may have listings of free mulch materials, or you can place your own listing requesting them. Just be sure that your material is weed-free and that it has not been treated with any chemicals that you would not want in your garden.

Mulching is an ongoing process because the material breaks down and turns into soil. I mulch year-round as necessary, keeping the mulch topped up to at least 2 inches in annual beds, 4 inches around perennials, and up to 12 inches on paths. Late fall applications are best for starting new beds, so the mulch has time to break down over the winter and prepare the soil for spring planting.

Tree mulch

I like the fresh smell of chipped leaves, twigs, branches, and trunks, and neighbors always comment on the wonderful scent that permeates the air when I mulch with these tree trimmings. I prefer to use material from living trees that have carbon-rich structural wood; these are shredded in a chipper and delivered by a local company. The resulting material has greens, browns, air, and water mixed together by the chipper. It takes a few years for this type of mulch to break down, but when it does, it becomes black, crumbly, rich soil ready to use. During the five years I've lived on my 6000-square-foot lot, I've covered the ground with about 85 yards of tree trimmings with the help of volunteers, fellow gardeners, interns, and students.

When you get tree mulch delivered, start with a 5-yard delivery. Note that you can request stumps to come in the loads to use for stump benches or to border raised beds. If your space is too small for a whole truckload, organize a few friends at a central location and then take smaller truckloads to individual gardens. The trimmings will usually be on the truck for a day or two before coming to your site, and you will feel the heat and see the steam rising from the pile that tells you the mulch has already started decomposing.

You need some tools and some help to move large quantities of mulch. I park a wheelbarrow next to the mulch pile and use a pitchfork to dig the mulch into the barrow, switching to a snow shovel when the pile gets closer to the ground. Finally, I use a broom to sweep up the remains.

In the garden, I put the first loads of mulch the farthest away from the delivery site, so I gradually decrease the distance I have to wheel the barrow. Spread all of the mulch in paths, not on the beds. A mulch layer of 6 to 12 inches starts compacting and breaking down to 4 to 6 inches within a

Dried straw—with no seeds—makes a sweet-smelling mulch that protects and enriches the soil food web.

month. It looks clean and smells fresh after a new load, and it's a pleasure to walk on the springy paths.

You can use tree trimmings to make moisture-retaining mulch pits for large perennials and trees. About 12 inches away from the planting hole, dig a 12-inch deep trench as big and wide as you can. Fill it with shredded paper, such as old telephone books, then cover with a thin layer of tree mulch. During dry weather, the roots will feed from the stored moisture inside the mulch pit.

When pruning trees and shrubs, practice the chop and drop method. Use ¾-inch or less diameter branches that will rapidly decompose. Save larger branches for fires or fencing. For softwood stems and leaves, simply return the prunings to the base of the shrub or tree where they were growing. Place hardwood trimmings along paths. Take the extra time to cut or chip them up into smaller pieces so they're easy to walk on. Don't use diseased trees because this might spread the disease more widely. For these, use your municipal composting program—but check with them to see what materials are acceptable. You can also use the prunings of herbaceous perennials—plants that die back in the winter and return the next spring—like comfrey, sunchokes, and yarrow. The cut leaves can be placed around other plants as mulch, or tucked under existing mulch to add green material.

Sheet mulching is a technique of covering the ground with layers of organic matter, much like layering a lasagna. It's an excellent way to start new beds over weedy or unproductive soil, or to replace an existing lawn. This no-till method saves work and allows the material to decompose over time, and you can plant into the mulch right away with big seeds and transplants.

TOP LEFT
Placing leaves in piles is a simple way of creating mulch. Keep the leaf pile moist (but not soggy), and after just a few weeks, the leaves will start to break down to make leaf mold.

TOP RIGHT
Woodchips make excellent mulch and can be used throughout the garden, both in the beds, along pathways, and in swales and drainage channels. Over time, the woodchips decompose and become soil.

BOTTOM LEFT
All kinds of creatures find shelter in a permaculture garden. Here a quail has laid her eggs in a sheltered spot amid leaves and small branches that have been chopped and dropped in place.

BOTTOM RIGHT
Sheet mulching can help prepare a large area for planting by smothering an existing lawn or weedy patch. Helpers make the job go much more quickly.

Sheet Mulching

Your sheet mulching ingredients will depend on what's locally available. To begin with you need cardboard or newspaper. Sources for large cardboard boxes are bike shops and furniture stores, but be sure to remove any staples and packing tape. Do not use the glossy paper from magazines. You will also need amendments, if they were recommended by a soil test. These might include blood meal, bone meal, feather meal, guano, grass clippings, rock dust, or seaweed. You need whatever animal manure is available—chicken, cow, goat, horse, rabbit, or sheep. Finally, you need other organic waste in the form of kitchen scraps, brewery waste, coffee grounds, juice pulp, municipal compost, old potting mix, leaves, straw, tree trimmings, woodchips, bark, or sawdust. Do not use plywood or wood that has been pressure-treated. Diversity of the layers is a critical component so try to get a good mix of greens and browns.

The key is to really smother the underlying material, so pile the materials deep. It is better to start with a small area and make it 12 inches deep, rather than sheet mulch a larger area and make it too thin.

You can vary the mix for specific types of crops like fruit trees, shrubs, perennials, or annuals. For annuals, use straw and horse manure for faster decomposition. For trees and shrubs, use woodchips, tree trimmings, and sawdust for a slower decomposition.

For best results, sheet mulch in the fall and plant in the spring. For planting a sheet-mulched area right away, finished compost can also be added to individual plantings like potatoes, tomatoes, squash, and other large transplants.

Instructions

1. Water the area to be mulched for several hours.
2. Cut down existing vegetation and leave the trimmings or grass clippings in place.
3. Spread appropriate amendments, if needed.
4. Put down 1 inch of cardboard or several layers of newspaper, overlapping each section.
5. Water the cardboard or paper thoroughly.
6. Add a 9-inch layer of manure and organic waste.
7. Water once more.
8. Cap with a 2-inch layer of bark, sawdust, straw, or woodchips.
9. To plant immediately, move back some mulch and make an X in the paper layer. Dig a planting hole, add a shovelful of finished compost, and set the plant or seed in the hole. Replace the mulch around the base of the plant.

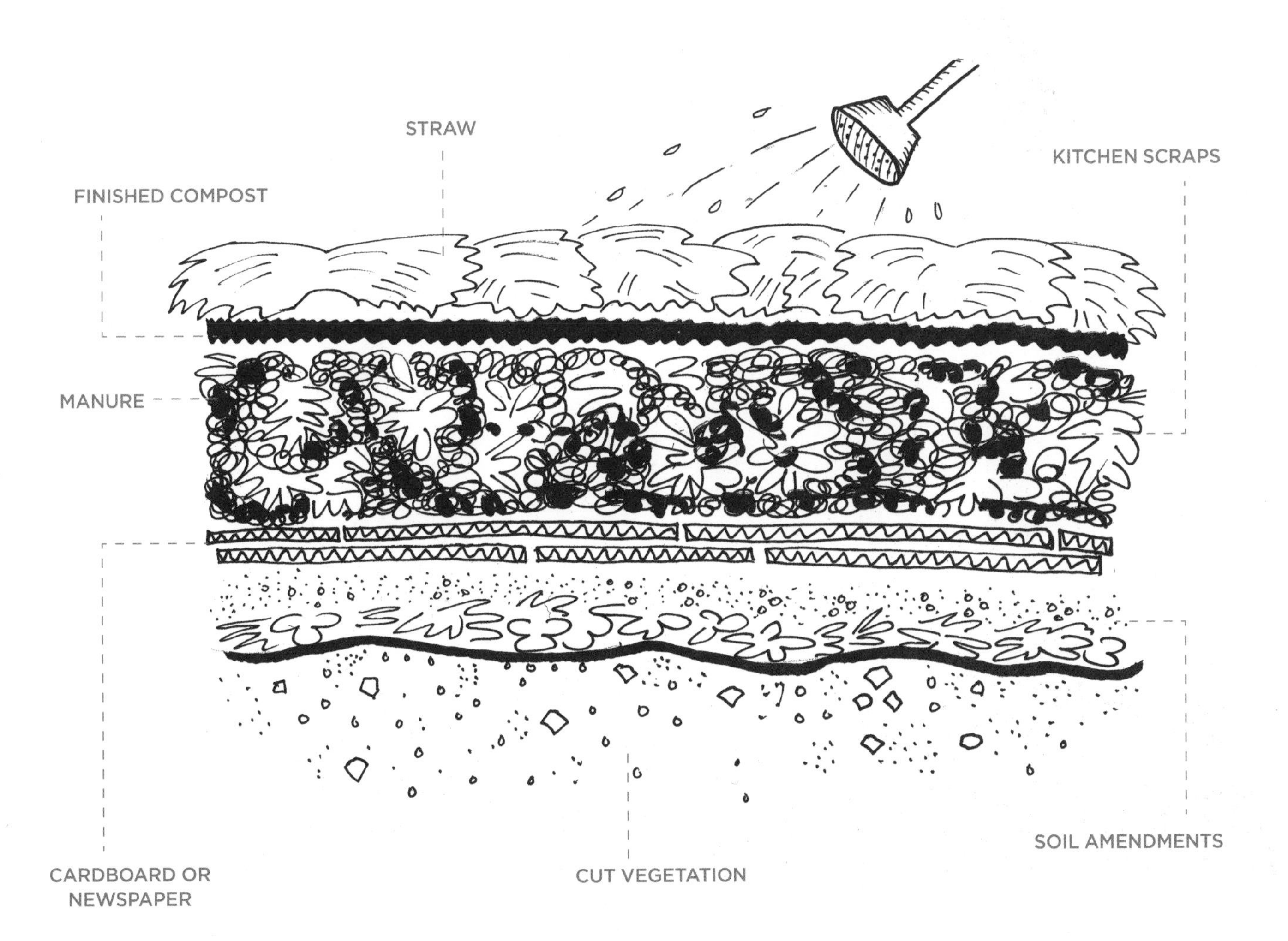

Cover Crops

Cover crops are planted, not to feed us, but for the soil to eat. They are nurse crops that provide nitrogen, carbon, other minerals, and organic matter to feed the soil food web. In addition to giving a boost to the next season's crops, they act as a living mulch, suppress weeds, and provide forage and habitat for beneficial insects. Because the entire biomass of the cover crop plant—roots, leaves, stems, fruits, flowers, and seeds—are dug into the ground, they improve soil tilth, structure, and water absorption. There are nitrogen-fixing cover crops and dynamic accumulators that are especially useful for adding nutrients and minerals.

Like sheet mulching, cover crops are good ways to start new garden beds. In late fall, plant cover crops like bell beans, vetch, and annual rye. In the spring, cut them down and dig them in or cover them up with mulch. Within a few weeks, decomposition and the release of nutrients will have occurred and you can plant this season's crops.

If you have a big garden, you may want to keep unused beds planted during the warm season with something that improves soil for the fall crop. I use buckwheat or beans for a summer cover crop, and dig in or cover up with mulch, wait two weeks and then plant squash. In colder climates, plant unused beds with cover crops in the summer to prepare for crops the following spring.

Cover crops like these fava beans are multifunctional plants. They produce edible seeds, fix nitrogen in the soil, attract beneficial insects, and add biomass when you cut them down or plow them into the soil.

SEASONAL COVER CROPS

COOL SEASON NITROGEN-FIXING	WARM SEASON NITROGEN- FIXING	COOL SEASON BIOMASS	WARM SEASON BIOMASS	PERENNIAL NITROGEN-FIXING	PERENNIALS BIOMASS
Austrian winter pea	soybean	barley	buckwheat	alfalfa	chicory
bell bean	black bean	Japanese millet	sudangrass	red clover	comfrey
crimson or sweet clover	cowpea	mustard	sunhemp	white Dutch cover	dandelion
fava bean	pinto bean	oats		white Ladino clover	nettles
fenugreek	sesbania	spring and winter wheat		white New Zealand clover	yarrow

Edible flowers and different plants from the brassica family happily mingle in permaculture beds.

5 PERMACULTURE EDIBLES

Choosing Crops

Permaculture is ecological gardening—the act of putting edible foods into an environment where they will produce, with inputs generated onsite or found locally. For the most satisfying experience and bountiful yields, you need to familiarize yourself with a range of annual and perennial edible plants, and know the basics of planting, tending, harvesting, processing, sharing, and seed saving. But remember that the goal is to let nature do the work for you as much as possible.

Vegetables and fruit are divided into various groups, and each has its own unique cultural, regional, and seasonal characteristics. When putting together guilds, it's helpful to think of similarities that edible plants may share, whether they belong to the same plant family, originate in a certain region, are seasonal partners, or are used in a similar way for cooking. There are many familiar choices as well as some unfamiliar ones you may want to try. I've written short descriptions of my favorites that are easy to grow across most of the country and really productive for a beginning gardener. Most can be eaten in-season or stored, and include a broad cultural palate.

Don't forget that in permaculture we want to observe and interact, so pay attention to local growing conditions. Most areas across the country are frost-free from June to September, but in some areas, planting as late as July may be necessary to avoid late frosts. In other areas, a few hardy crops like lettuce, onion, radishes, and spinach can be planted as early as April.

Consult with your neighbors and friends to see what grows best in your area, and find out what's growing in your neighborhood

A summer harvest of mixed crops includes tree fruit, berries, root crops like carrots and beets, vine crops like zucchini, and heat-loving plants like peppers and tomatoes.

and local community gardens. Next, check out your local county cooperative extension office, and Master Gardener groups for classes, events, and other nearby knowledgeable experts. You can also consult with local nurseries to see what thrives in your area and when to sow or plant it. This kind of information is especially helpful if you aren't familiar with the frost dates in your area, or the number of chilling hours, which are the number of consecutive winter hours below 45° F. Where I live in the coastal Bay Area, chill hours usually add up to about 400, but just a little inland it goes up to 900. Some varieties of apples need as much as 1800 hours, so those varieties will never do well here.

The permaculture principle to use and value diversity encourages us to try out many new things in our gardens. We need to learn from our mistakes and build from our successes to adapt to our particular habitat and bioregion. Every year the growing conditions will be slightly different. One year a cold snap will come in and kill a tender seedling, and the next spring, hail knocks flowers from certain fruits. We need to cultivate an attitude of playful experimentation and have enough crop diversity to be resilient when things inevitably go wrong, yet still get enough to eat.

Fruits and Nuts

My older daughter loves to eat fresh fruit in the garden. When she was little I had to remind her not to pick berries when they are green or white. She has learned to wait and pick strawberries and raspberries when they're dark red, and blueberries when they are all blue. I love the fact that picking, sampling, and arranging fruit can keep her occupied for long stretches of time. Now that I have a growing family, I am planting even more fruit to keep up with the demand. It's deeply satisfying to have seasonal fruit I grow myself to share with family and friends when they come over.

Climate and taste preferences will of course affect your choices of fruit trees, shrubs, and vines. If you live in Florida or California, you might grow Meyer lemons or pomegranates, but in Michigan or New York, chances are that you can find an heirloom apple variety or a locally developed grapevine that thrives in your region. Remember always to evaluate and respond to change. When you find a plant that does well and you enjoy eating, keep growing it. If your blueberries are thriving, plant more next season. Consider adding a nut tree if you have the space. Continue to observe and experiment over time.

This principle of observation also applies to other creatures that are drawn to your edible garden. Given how tasty these crops are, it's no surprise that furry, feathered, and flying pests also want to eat them, so you may need to use some strategies and techniques to get rid of unwelcome foragers.

SHARING THE HARVEST

Fruit crops are particularly good for practicing the permaculture ethic of sharing, because their abundance during the peak harvest season is generally too much for the average family. Where I live, there are many wild plums, also known as cherry plums because of their small size. The seeds are spread all over town by birds and squirrels. When I first moved into my current house, there was a huge wild plum tree in the backyard so we had lots of fresh plums for ourselves and to give away to neighbors. One night I stayed up late making plum jam and ended up with 24 quarts!

In my experience, people with fruit are happy to trade and share. Most neighborhoods have treasures of unharvested fruit, and with a little polite investigation and networking, you may be able to gather some of the harvest for yourself.

The agricultural past of many communities may be reflected in place names like the Pear Tree Shopping Mall in Palo Alto or the Fruitvale district in Oakland. In Moraga, California, there are remnants of old pear orchards, and people still gather up the summer bounty from the gnarled old trees. In Davis, California, there is a street named Olive Drive where people gather olives for salt or brine curing. There may also be yard-share or fruit-share organizations in your community.

In other places, fruit trees are widely planted as ornamentals. Where I grew up in Michigan, crabapples are common landscaping plants because of their beautiful flowers and classic shade tree form, but the fruit tends to go unharvested. My mother and I would gather crabapples from our neighbors' trees and make jelly. Gardening is all about timing, and beginning gardeners often have to learn what crops look like when they're ready to be harvested and eaten. Fruit on city streets may get picked before it is ripe and then tossed after a single bite. Neighbors of mine have lost unripe apples, plums, and peaches from their front yard to people helping themselves to fruit. Now they put up a sign that says, "Not ready! We will put out a free box when it's ready."

Another option that aligns with the permaculture principle of fair share is gleaning—getting permission from a farmer to collect remaining fruit after the main crop is harvested. A few years ago at an olive orchard near my home, the farmer asked people to come out and help him pick olives. About ten people came out and spent a very full day picking olives. We harvested about a ton of fruit, and when it was pressed in a local facility, we each got a little over a gallon of olive oil.

FRUIT TREES

Fruit trees are a study in patience and practice. A fruit tree is the central element in a fruit tree guild because it's the largest plant and calls for the most planning. In a new garden, it's helpful to place semi-dwarf or dwarf trees first and design around them. If you have existing fruit trees, it's possible to build around them, but it requires you work with the existing conditions in your space. For instance, if you have a mature standard tree, it's hard to plant under it because those plants won't get any sun. Older trees can be gradually pruned down to around 8 feet over a three-year period to allow more space for sun-loving plants below. You can also thin out the branches, which reduces the risk of diseases by allowing greater air circulation.

Qualities to look for when selecting varieties of fruit trees include hardiness, the number of chilling hours required, disease resistance, and pollination needs. Some trees self-pollinate; others require a pollinator. Always ask about the pollination needs of trees; a good nursery will provide this information. If you don't have the room for several varieties, you may be able to benefit from a neighbor's tree. Even cutting flowering branches from a suitable pollinator

Fruit trees form the main canopy layer in the food forest. In the center of my garden, I have apple trees, apricots, and tree tomatoes—some of my favorite food crops.

Wildheart Gardens

Bill Merrill maintains forty-eight varieties of grafted apples on just twelve trees in his Fremont, California, garden.

OPPOSITE
I plant dwarf and semi-dwarf apples in my garden because I want to keep them as part of the small tree layer. This one is underplanted with asparagus.

and placing them in a pot beside your tree can do the trick, as long as you have helpful insects like bees nearby. In small gardens, you can plant a multi-variety tree that has two or three different scions grafted onto a single rootstock.

Fruit trees can produce a big crop in a short time, so be prepared for the overflow. It takes a lot of time to deal with the abundance of a crop, so plan ahead for what you will eat, process, or share. Of course, eat as much fresh fruit as you can because it's so good for you and tastes best freshly picked. Preserving the harvest by drying, canning, fermenting, and freezing for use during the non-fruiting season extends your choice of nutritious fruit for the full year. Then think of ways to diversify what you have by trading it with family, friends, and neighbors. Or practice fair share, and drop off boxes of fruit to schools, food banks, and churches that feed the hungry.

POME FRUITS

This category includes both familiar fruits like apples and pears, as well as some uncommon ones, like medlar and quince.

Apples

Apples are probably the most popular fruit in America; they can be eaten fresh or dried, squeezed into juice, fermented into cider, stewed, and baked into pies and cakes. I particularly love their spring blossoms, which signal the coming of a new season and attract bees to the garden. Most people are familiar with Red and Golden Delicious, Fuji, Gala, and Granny Smith, but there are hundreds of apple varieties with different colors, flavors, ripening times, and storage abilities.

Apples do not appreciate extremely high temperatures, so they aren't suitable for the hottest regions. Because different varieties ripen at different times from summer through late fall, you can plant several varieties to provide fruit for many months of the year. Varieties that are resistant to challenging diseases like apple scab, fireblight, and crown rot are also available. Apples are grafted onto rootstocks that control the overall height and spread of the plant; look for semi-dwarf types that grow to no more than 15 feet. In small gardens, try columnar varieties, which are good space-savers and are perfect for containers.

Pears

Pears can be long-lived and often are more resistant to pests than apples. They make a wonderful dessert fruit, and are almost as good when canned. If you have clay soil, pears are a good choice as they can tolerate heavy soils better than most fruit. European pears need more winter chill and lower

average temperatures; if you live in a hot area like California or Florida, Asian varieties may be the best choice. Pear trees can grow big, so they are usually grafted onto quince rootstock to limit their height.

Other pome fruit

Quince are beautiful little trees that fit well into the permaculture garden. The fruit of most varieties cannot be eaten fresh, but makes wonderful jelly, chutney, or jam. They require fewer chilling hours than other pome fruits and are often long lived.

Loquat is an evergreen warm-climate fruit tree that you can try if you have the right climate and the space. This is one of those fruits that you don't often find in supermarkets because the fruit doesn't store or ship well. It's not a big tree—growing only to 15 feet, but you need at least two trees to ensure pollination. It's a good low-maintenance choice. The fruit may be white, yellow, or orange, depending on the variety, and is succulent and tangy.

Medlar is a traditional European fruit tree that is quite rare in North America. The fruit looks a bit like a persimmon and is picked unripe, then stored for several weeks to eat once the pulp becomes squishy (a ripening process known as bletting).

STONE FRUIT: APRICOTS, CHERRIES, NECTARINES, PEACHES, AND PLUMS

These are the quintessential, super sweet late spring and summer treats. Called stone fruits because of the pit in the center of the fruit, they all have delicate aromas, high sugars, and tender flesh. Growing stone fruits is highly dependent on climates, microclimates, and soil. In northern gardens, you'll want to choose varieties that are winter hardy and bloom late to avoid spring frosts that can reduce pollinating insects and damage early fruit. In moist climates such as the Pacific Northwest and South, fungal diseases can be a problem. In the warmest climates, you need to be sure you have sufficient winter chilling hours.

Cherry

There are many different species of cherry but the two main types for gardens are sweet and sour, or pie, cherries. Be sure you know which kind you are planting; sour cherries are edible but are preferred for cooking and drying. Sweet cherries tend to be better suited to warmer climates. Most cherry trees require a pollinator. Cherries are grafted onto rootstocks that make them suitable for a wide range of climates and conditions; be sure to choose one that limits the size of the plant, as cherry trees can grow to 40 feet in time.

Plums

If you are planning to grow plums, you have even more choices. There are Asian types, European plums, and crosses between the two. They can be eaten fresh, dried as prunes, made into jams and jelly, or even fermented into wine. Again, check the pollination needs and rootstocks before you choose a variety.

In some regions you will find wild plums or cherries, and these are treats to be relished. Some skilled grafters have been successful in using local native plums and cherries as rootstock to grow named varieties on, in locations that would not normally be suitable for regular rootstock (such as in wet soils).

Quince trees produce yellow fruits that are something like a cross between an apple and a pear. The fruit ripens from late summer through early winter.

Apricot

One of my favorite stone fruits is the tangy and sweet apricot. Apricots are the exception to the rule of most other deciduous fruit trees; to avoid diseases, they are pruned in summer rather than winter. These are a shorter-lived fruit tree than apples, and on a commercial scale, are cut down and replanted on a twenty-year cycle, so keep that in mind when you have an older tree and are wondering why it's not doing so well. Sort harvests into fresh eating fruit that will last more than a week, with the remainder left for cooking, drying, and canning.

Citrus: grapefruit, lemons, limes, and oranges

Citrus are warm-weather trees and certainly don't thrive everywhere. In colder climates, you can grow citrus in containers or in greenhouses. Even in warm climates, be sure to seek out types and varieties that are suitable in your area. You may need to plant your citrus trees against a south-facing wall, or protect them from frosts by covering with sheets or row covers. Mandarins are the most cold-hardy of citrus.

Meyer lemon is my favorite citrus because it's a great producer in cool summer climates like mine and it has a sweeter taste than the average lemon. It's fairly pest free and you can harvest fruit for most of the year. I use Meyer lemons to mix into lemonade, squeeze onto salad greens, make lemon bars or zest for pasta, or as preserves.

TREE TOMATO

Tree tomatoes (*Cyphomandra betacea*) are unusual subtropical fruits that are members of the Solanaceae family, like tomatoes, peppers, and potatoes. They are fast-growing trees that reach about 20 feet in height. The fruit resembles tomatoes or tomatillos and can be used in similar ways—made into soups or pasta dishes, or put into salads.

OTHER: AVOCADOS, FIGS, PERSIMMONS, POMEGRANATES, OLIVES, AND MULBERRIES

Food crops go through trends just like fashion. It wasn't that long ago that pomegranates seemed exotic, and now most people are familiar with the nutritional benefits of pomegranate juice. In our market-based agricultural system. farmers are concerned with mass producing crops for large stores so they plant fruit like mulberries to distract the birds instead of as a food crop. However, the permaculture gardener isn't limited by the need for mass production and can grow delicious crops like mulberries on a smaller scale. You will, however, need to provide some protection from the birds.

Figs are a permaculture favorite in warmer regions (although they can survive

RIGHT
This Meyer lemon thrives in my cool-summer garden, producing fruit almost all year long.

TOP AND BOTTOM
Tree tomatoes prefer a sheltered site and need lots of mulch to keep the soil moist and reduce drought stress. The fruits ripen from green to red over a long period so we harvest them regularly.

winter temperatures to −10° F), fast-growing, with a spreading growth habit. They can even be trained over a pergola or arbor, and in some climates will produce two crops a year.

Pomegranates have showy red flowers and naturally form bushy rounded plants up to 20 feet tall, but they can be pruned up into more of a tree form. They need a hot climate, so in marginal areas, they are best grown against a south-facing wall to help ripen the fruit in fall. You can also grow pomegranates in large containers.

Persimmons are also a fantastic permaculture crop because they give fruit late in the year (November–December). There is no other fruit that matches persimmons for sweetness in the winter. Persimmons are tropical-looking trees with glossy dark green leaves and attractive orange fruits that hang on like ornaments after the leaves fall. They can take part shade when young which makes them a useful understory tree. Most persimmons are astringent if eaten fresh and therefore are ripened off the tree until soft. I have gleaned persimmons, dried some, and also filled my freezer with them, enjoying them the next summer in fruit smoothies or as a do-nothing ice cream—just take them out and eat as they soften. 'Fuyu' is a non-astringent type that can be eaten like an apple, although the flavor improves if you let it soften.

Avocados can only be grown in frost-free areas. These are some of the best-tasting and healthiest foods available, with omega-3 fatty acids that are rarely found in plants. They have an A type and a B type whose flowers open either in the morning or afternoon, and you will need two compatible trees for pollination. Avocados are huge trees, up to 30 feet tall and wide, and need lots of feeding during the wintertime. They die or struggle if they are in a poor drainage situation. Harvest the fruits when they are still hard and ripen them off the tree.

Christopher's Garden: FAVORITE FRUITS

My birthday is in the winter and my favorite birthday cake is persimmon cake made with acorn flour. This is slow food at its best, where I set aside some special time to gather the acorns. My older daughter likes to help crack open the acorns and grind the flour. When she was born, my mother gave us a 'Hachiya' persimmon tree because it's my favorite winter fruit. Now, four years later, I'm looking forward to some persimmons from my own garden this winter.

NUT TREES

Because most nut trees are so large, they are generally part of the canopy layer, but there are some rare dwarf types such as the filbert. As with large fruit trees, it's really important to think about the size of your garden when planting nut trees or planting around them.

Generations before us understood the importance of planning for the future. For example, one hundred and fifty years ago, whenever a church was built, an oak tree was also planted to benefit future generations when they needed the wood to do repairs on the church. We take the same approach in permaculture, thinking about how we can plant renewable resources for ourselves and for those who will inherit our gardens.

LEFT
Olives are beautiful landscaping trees and can grow for many years. They produce fruit after two or three years. In cold-winter climates, you can grow them in large containers and bring them indoors in winter.

RIGHT
This young almond tree in my garden forms the center of a fruit tree guild, surrounded by chard, herbs, and a generous mulch of straw.

ABOVE
When almonds ripen, the outer hull starts to crack open. Knock or shake them off the tree onto tarps, peel off the hulls, and spread the nuts in the sun to dry.

OPPOSITE
I harvest nuts in my own garden, in the gardens of friends, and in the wild. These are black walnuts (TOP) *and acorns and baynuts* (BOTTOM).

With nut trees you also need to think about how long it will be before the tree produces a yield. Some nut trees produce in a relatively short time. Almonds can produce in as little as two to four years. Others, like stone pines, can take five to ten years to start producing and forty years to reach their full-yield potential.

Common favorites are almond, filbert, and walnut. Less common are bunya-bunya (*Araucaria bidwillii*), butternut, chestnut, hickory, macadamia, oak, pecan, pinyon pine, and stone pine. There is no one nut tree that will work well across the country, so where you live determines which nut tree produces best. Check with your local cooperative extension office, go to farmers' markets, and ask other growers what is being grown in your region.

Filberts are good garden-scale nut trees, as they grow from 10 to 18 feet in cold climates. I've seen them growing on the coast from northern California to British Columbia—areas with mild, moist winters and cool summers where they get sufficient chill hours.

California has a good climate for almonds, and this is where most of them are grown commercially. While almonds are closely related to peaches, they need hot summers and can be tricky to grow in winter-freezing areas. 'All-In-One' almond is my favorite nut to grow. It is self-fruitful and low-chill but winter-hardy. It generally needs hot summers to ripen, but bears heavily in my coastal growing area where there does seem to be enough summer heat for it to bear nuts. The soft-shell nuts pop open when they are ripe. The nutmeat is sweet and has a delicious flavor. At 15 feet, it's small for a nut tree, making it a good choice in smaller gardens.

If you have the space, walnuts are a nutritious and cherished crop; however, they grow 30 feet tall or more and almost as wide.

I love to eat walnuts, but one tree would take up most of my yard and their roots are allopathic, which means they can be harmful to surrounding plants. My neighbors have wild native black walnuts in their backyard, but I'm still trying to beat the squirrels to the harvest, and I usually don't get many because the trees are so big they aren't easy to harvest.

While American chestnuts have become mostly extinct due to a disease, Chinese chestnuts grow well in most parts of the country with enough winter chill and hot summers. Chinese chestnuts can grow to 45 feet tall and wide. All chestnuts can be coppiced, or cut down, every few years and the wood used to make fences and posts, but then you wouldn't get any nuts.

Pecans are native to the south central United States, and are grown in warmer parts of California, too. They are massive—about 70 feet tall—so you would need a very large yard. Shagbark hickory is a possibility for very cold climates in the Midwest.

I think oaks are highly underrated as a food source. They grow all over the world, and wherever oaks have grown, people have eaten them. Newer varieties of white oak have been selected to have the least tannin, the substance that gives the nuts a very bitter flavor but that also can be leached out by placing the nuts in running water or repeatedly boiling them. I leach the ground nut meal by placing a small amount in a coffee filter and rinsing until the water no longer turns dark brown.

I gather local coastal live oak acorns in the hills every year and make breads and crackers, but it takes some work to crack, grind, and leach the acorns. The taste is worth it, and it's very satisfying to connect to a real local indigenous slow food

tradition. This is one tree to consider as an abundant wild crop in the forest rather than bringing it into the garden, unless you have a large property.

Wherever pine trees grow, edible nuts can be found, but not all pines have cones with seeds large enough to harvest. For the arid Southwest, pinyon pines (*Pinus edulis*, *P. monophylla*, and *P. cembroides*) are the best species for nuts. Other pine trees with edible nuts include gray pine (*P. sabineana*), Korean pine (*P. koreana*), stone pine (*P. pinea*) sugar pine (*P. lambertiana*), and Torrey pine (*P. torreyana*).

Ginkgo can be grown widely and is not a popular edible nut tree, but I have eaten the nuts locally in the Bay Area. Unfortunately, the female tree is rarely planted as the fruit has a very unpleasant smell, even if the nuts taste good. Other nut trees for warmer climates are monkey-puzzle tree, pistachio, macadamia, and Chilean wine palm.

Other nuts

There are other edibles called nuts that do not actually come from nut trees. Peanuts are nitrogen-fixing perennial legumes that grow well in areas with a long growing season and light, sandy soil. George Washington Carver helped revive agriculture in the South by promoting these as a rotation crop to improve the soil. Sunflowers are grown for their edible seeds and brighten up any garden. Like other members of the family Asteraceae (formerly Compositae), they are magnets for beneficial insects. The trick is to beat the squirrels and birds to the seeds, so after the flowers start to fall off and the seeds are ripening, you can tie a brown paper bag over the head to make it less appealing to these creatures.

Think beyond trees when you are looking for protein-rich food crops for the garden. Peanuts are part of a legume plant that develops underground. Sunflowers produce hundreds of edible seeds encased in shells.

CARING FOR FRUIT AND NUT TREES

In the North, the best time of year to plant fruit and nut trees is in early spring, so that spring rains can help the roots to get established. In milder climates you can plant in fall or even winter. Either way, you should do your soil preparation the year before if possible, by adding compost, mulch, and a cover crop.

To have success from your trees and shrubs, you have to start with really healthy nursery stock. Dormant, bareroot deciduous trees are best to plant in early spring, but in some areas, or for citrus and other evergreens, you will have to plant from container stock. If buying a container-grown tree, remove the plant from the pot and examine the roots. Don't buy rootbound plants, or those with brown roots—the roots should be healthy and white.

It's more of a challenge to dig a planting hole for a tree in clay soil, and you must be sure the hole drains sufficiently. Some trees are especially susceptible to poor drainage, including avocados, citrus, cherries, peaches, and walnuts—look for rootstocks that can tolerate clay soil. Dig the hole bigger than the plant's rootball. If the sides of the planting hole are slick surfaces, the tree will remain rootbound. Take a digging fork to score and break up the sides and bottom of the hole, making channels for roots to forage for water and nutrients.

I don't recommend amending a planting hole, but prefer to use the native soil to fill in around the tree's roots. Create a little mound of soil in the base of the hole and gently drape the roots over it when planting. Place the plant a little higher than the soil around the hole, to prevent crown rot and to leave space for the mulch layer (unless you are in an arid climate and planting into a mulch pit).

I mix together some of the native soil with compost and form a doughnut shape around the base of the tree, about a foot from the trunk. This creates a little pool for giving the tree a good soak when first watering it, and keeps the compost away from the base of the tree. Over the next few years you can make the pool wider and wider if ongoing watering is needed. If you are planting on a slope, place bricks, urbanite (recycled broken concrete), or small logs uphill in a semicircle to keep compost and mulch from slowly migrating downhill and piling up at the base of your shrub or tree.

Small fruit trees and berry bushes can be irresistible to birds. You may need to cover the plants with nets so that you don't lose your entire harvest.

PESTS

Just as humans enjoy the garden's bounty, so do many animals, birds, and insects. Most gardeners have learned to think of pests as enemies of the garden, but the permaculture approach is first to accept that a certain percentage of your crop will be donated to the worthy cause of biodiversity. A healthy, diverse garden attracts beneficial creatures like parasitic wasps, frogs, lizards, and snakes that will do most of the natural pest control. You can attract these helpers by creating habitats such as a pond (for frogs), a rock pile (for lizards), or a pollination hedge (for beneficial insects). Finally, there are a variety of techniques you can employ to keep pests at bay, from scarecrows to row covers to hand-picking.

Moth larvae, aphids, mites, and fruit flies are all attracted to fruit. Pheromone traps and sticky traps can reduce their numbers. Banding tree trunks with sticky materials and covering burrs and tree wounds with clay-based compounds (kaolin) can prevent pests from laying eggs. Many tree pests drop to the ground to pupate. Be sure to clean up debris and fallen fruit from the ground underneath trees to discourage pest reproduction. If you have chickens, encourage them to peck around and underneath trees to consume insect larvae. If you have a visible population of beetles, such as weevils or curculios, shake the tree vigorously so that the beetles drop to the ground into a tarp or the waiting beak of a hungry chicken.

Fruit-eating birds can make short work of your crop. Cherries are especially susceptible to bird predation, but I've had good luck planting sour cherries (or pie cherries)

that birds don't seem to attack. Some farmers plant mulberries as a sacrifice crop for birds to eat. Methods to scare away the birds include tying reflective items like old CDs and silver tape among the branches, or erecting decoy models of hawks and owls at a prominent height. You can be creative and design your own scarecrows. Netting is somewhat effective, but it can be a pain to put up each year and makes harvesting harder; if birds are a persistent problem you may have to install permanent netting over a giant wooden frame several feet higher than the maximum height of the trees.

Four-legged creatures like squirrels, chipmunks, and raccoons can be tenacious and persistent pests. In many places, deer are persistent and incredibly destructive, browsing on both edible and ornamental plants. Some adventurous gardeners take the permaculture principle to produce no waste literally, and instead of seeing pests, see an opportunity to trap or hunt some dinner for themselves or their pets. For those not ready to take that step, there are other solutions.

If you live in the country or near wild lands, natural predators like coyotes, wolves, mountain lions, and snakes can help keep populations of squirrels and deer in check. In more residential settings, the family cat or dog can deter rodents (although a hunting cat is a danger to birds as well). Trapping can be effective for rats and raccoons, but the traps can pose a threat to children and pets. So-called humane traps are not encouraged in permaculture, as catching and releasing pests elsewhere just creates the problem for others.

PRUNING

In the permaculture garden, it's best to prune fruit trees so the fruit is accessible from the ground. Dave Wilson Nursery, possibly the largest national grower of fruit trees, recommends pruning fruit trees to only 6 to 8 feet and keeping them that size. The best time to do this kind of containment pruning is during the summer, after the tree has fruited. Summer pruning slows the overall growth of the tree and can enhance fruit bud production. Winter pruning is done when trees are dormant and it's easier to see the entire structure and take out dead or diseased wood and crossed branches. Winter pruning stimulates growth of the fruit tree the following season.

Another technique for limiting the size of fruit trees is espaliering, growing a tree flat against the surface of a wall or fence. If you have a south- or west-facing wall, this can also be a way in marginal climates to grow heat-loving crops like citrus or fig. You can grow many dwarf fruit trees in containers, but every few years you will have to take out the plants, prune the roots, and repot the tree.

Not only each kind of fruit or nut tree, but often each variety, is pruned differently. For example, a white and a black fig are pruned differently (white figs produce fruit on last year's growth and black figs from current season's growth), so it's important to do your homework. Some trees, like persimmon and citrus, don't need much pruning. Others will not fruit without it. It's often a good idea to get help from someone who knows what they're doing the first time around.

GRAFTING

Grafting is the technique of taking wood from an existing tree with desirable fruiting characteristics and combining it with a desirable rootstock so that the two grow together to create a new plant. This is why we still have apples like 'Spitzenberg' in production that were first enjoyed hundreds of years ago. Grafting takes practice,

It takes a bit of time to train fruit trees into an espalier, but it's a good technique for maximizing space and capturing heat.

and you may want to observe an experienced orchardist or nurseryperson to see their technique. You can obtain rootstock from a tree nursery or you can graft onto a growing tree.

To make a new clone of an existing tree, remove a pencil-sized piece of scion (budded stem) in the winter or early spring, depending on your last frost date. Use a grafting punch if you don't want to practice with a grafting knife. Select the scion to be the same diameter as the rootstock and carefully cut both pieces with the punch. Then push the scion into the keyhole of the cut rootstock, making sure the cambium layer (the thin, bright green outer layer that covers the inner wood) is touching on both scion and rootstock. Wrap Parafilm (found in grocery stores) around the graft and secure with rubber bands, then put a paper bag over the Parafilm and seal it. This prevents your graft from drying out and getting sunburned. Make sure to regularly water your grafted rootstock.

Check the graft periodically and when you see buds emerge from the scion, remove the bag. Wait another six months before removing the Parafilm and rubber

Grafting is an old technique that rewards the patient gardener with a greater variety of fruit and nut trees that are well adapted to local conditions.

bands. Then your grafted tree should be ready to plant in the ground or to repot in a bigger container.

LAYERING

Layering is a propagation technique that involves taking a flexible growing shoot or branch from a plant and burying a portion of it in a growing medium—typically soil—while leaving it attached to the parent plant. The shoot will form roots where it comes into contact with the soil and can eventually be detached from the parent and transplanted elsewhere.

Layering is typically done with woody species like fruit tree rootstock. You can also use the technique with woody perennials like rosemary or lavender. Blackberries and some raspberries can be propagated by tip layering, where the tip of a current season's branch is buried in the soil.

The technique is fairly simple. Remove any leaves from the underside of the shoot at the point where you intend to bury it. Bury the shoot to a depth of several inches and secure it with either a U-shaped stake or a rock. After anywhere from six to twelve months the new plant will have rooted and can be detached from the parent plant.

Cape gooseberries, also called goldenberries, groundcherries, and poha, originate in the Andes. The fruits grow on perennial plants that reach about 3 feet tall, or they can be trellised. They are carefree plants but will not survive frost.

Fruiting Shrubs

It's familiar landscaping tradition: a suburban home with a foundation surrounded by ornamental shrubs. These are often planted for privacy, but they bring little else of value to the garden. If you have some of these foundation shrubs around your house, why not take a step in the permaculture direction and replace them with shrubs that offer food as well as flowers, fall color, and screening?

Some common food plants for the shrub layer are blueberry, currant, elderberry, and gooseberry. Some uncommon shrubs to try are goumi (*Elaeagnus multiflora*), goji (*Lycium barbarum*), huckleberry (*Vaccinium* species), cape gooseberry (*Physalis peruviana*), seaberry (*Hippophae rhamnoides*), serviceberry (*Amelanchier* species), and tree tomato (*Cyphomandra betacea*). When you are integrating edible plants into the food forest, it can sometimes be tricky to determine whether a plant is technically a shrub or a vine, but that isn't the most important issue. Plant what you like and what will grow best in the situation. If you have a large garden, you might also plant native fruiting shrubs in the outer zones of the garden to satisfy the deer and the birds, keeping the cultivated types in zones 2 or 3 so they can be easily harvested (and netted, if need be).

Blueberries

Visitors to my home garden are often surprised to see blueberries, traditionally thought of as an Eastern crop. I've had good luck with the southern highbush varieties 'Misty', 'O'Neal', 'Sharpblue', and 'Southmoon', and I like that blueberries are also attractive landscaping plants with interest for most seasons. In the spring, they are covered with small pink urn-shaped flowers, they bear fruit in the summer, and in fall many varieties turn into deep red or yellow. Blueberries taste delicious, are packed full of nutrition, and are very productive plants. They do attract robins and other berry-feeding birds, so they are definitely a candidate for netting.

If I lived in an area with sufficient winter chill to grow northern highbush blueberries, I would choose 'Chandler', possibly the world's largest blueberry—it can grow 5 to 7 feet tall and is high-yielding. As a pollinator for 'Chandler', try 'Rubel', a classic heirloom northern highbush that fruits in midseason and has an outstanding flavor that is very close to that of the wild blueberry.

Currants and gooseberry

Black, pink, red, and white currants and gooseberries are all members of the genus *Ribes*. These plants were once widely grown in North America, but they fell out of favor because they are an alternate host for white pine blister rust, a serious pine tree disease. In some states, there are still restrictions against growing cultivated *Ribes* plants, but they deserve to make a comeback as there are new resistant varieties, and the fruits can be used in such a variety of ways: cordials, jams, jellies, teas, and wine. And they are perfect for permaculture gardens because they will bear fruit in the shade, making them ideal members of the food forest.

'O'Neal' is one of my favorite blueberry varieties and I grow bushes in zone 2, so that they can easily be harvested.

In general, currants and gooseberries will grow best in milder climates. Most require little maintenance, although the different types of currant bear on either two-, three-, or four-year-old wood, so be sure to verify the pruning needs of your variety before cutting back the plant in winter. Even if you only grow one of each bush, you will still be able to harvest enough fruit to make jam. *Ribes* plants can be propagated by cuttings and they may self-layer. Give the plants a dressing of compost every year.

Goji

You may have seen dried goji berries (*Lycium barbarum*) in the grocery or health food store. Goji originates in Mongolia and has been known for many years in Asia, but has recently been touted as a superfood in the West because of its high protein, mineral, and vitamin content. The berries grow on a sprawling shrubby vine that reaches up to 9 feet. If you don't have room in the vine layer, it can be trellised or pruned to a smaller shrubby shape. I love the flexibility of this kind of plant, which can be grown in a container, tolerates extreme low and high temperatures, and is mostly pest free. The fresh berries remind me of Jelly Belly candies, but you can dry them and then put the dried berries on cereal, mix them in smoothies, or bake them in muffins. The leaves of the goji can be made into a medicinal tea.

Goumi

Goumi (*Elaeagnus multiflora*) is also known as cherry goumi because of its small, bright scarlet, silver-flecked fruit. This is one of those special multifunctional plants that permaculture designers really like. It's an ornamental deciduous bush that grows to 6

OPPOSITE
Redcurrants grow on long string-like racemes and will ripen even in a semi-shaded understory. The fruit is bright red and sharp-tasting; you can use the berries to make jelly, juice, or syrup.

LEFT
The goji is a beautiful landscape plant that produces trumpet-shaped lavender- and cream-colored flowers that are pollinated by bees. They develop into small, bright red berries that can be dried and stored.

feet tall, with leaves that have silvery undersides and are dark green above. Goumi is a tough pioneer plant that can grow in a wide range of soil types, is drought tolerant, fruits prolifically, and fixes nitrogen in the soil. You can use it in the garden for screening, windbreaks, and as a distraction crop for birds to keep them away from your cherry tree.

When ripe, the fruit has a sweet yet tart flavor, and it contains vitamins and essential fatty acids, which is rare in berries. You can eat the fruit fresh, make jams, or dry the fruit. Goumi is partially self-fertile but produces better with a pollinator.

Vines

Vines constitute the most flexible layer in the food forest. These climbing or sprawling plants offer plenty of opportunity to use edges and value the marginal, and provide a way to layer edibles in otherwise wasted vertical spaces. Look around a forest and you will see vines sprawling up other trees and cascading down to take advantage of available sunlight. Nature does it, and so should we.

Vines can help provide privacy between structures, or between a building and the street. You can train them to cover arbors

Grapevines are vigorous deciduous plants perfectly suited to the vine layer of the permaculture garden. Grow them over a sturdy wooden arbor as they are long-lived and very heavy when loaded with fruit.

and decks, secure wires across a wall or a fence to act as a trellis, or even allow them to climb up a building and cover the roof. Vines can also cool a building or shade an outdoor structure from intense summer heat. In winter, deciduous species drop their leaves, which can be used as mulch, and the bare framework of the vine allows winter sunlight to filter through. Some vines can grow quite rampantly, so develop a realistic vision of how you want to integrate them into your garden and make sure you have a maintenance plan.

Christopher's Garden:

HEIRLOOM GRAPES

At Merritt College, we grow the local heirloom grape 'Emeryville Pink'. This unusual American grape was discovered growing near San Francisco Bay and unlike other cool-season grapes, it becomes sweet when it is ripe. A little over a hundred years ago all of the East Bay was farmland. Now Emeryville, which is a small city bordered by Oakland and Berkeley, is covered with houses and shopping malls. It seems a distant place from food production, but almost anyone in Emeryville could manage to grow one of these grapes, reconnecting the garden with the area's agricultural past.

Grapes

Grapevines are long-lived, multipurpose plants that can cover any vertical surface, whether arbors, fences, trellises, or walls. In addition to the fruit, which can be eaten fresh, dried, juiced, or fermented, you also can eat the leaves by pickling them for preparation as *dolmas*. Grapevines have other permaculture benefits. The fruit attracts beneficial wasps, and the fallen foliage and the pomace left over from winemaking make excellent additions to the compost. The canopy provides shade and fall color, and the ornamental twisted trunks are a sculptural garden feature.

You will need to select a variety best suited to your area and your needs. Grapes for winemaking are smaller and seeded, whereas dessert grapes are larger and often seedless. The range of grapevines and rootstocks is huge, so there are options for almost every garden situation.

The main task with grapes is to prune them correctly so that they bear fruit. Pruning grapes can be a complicated affair, but the important thing to remember is that they fruit on the previous year's growth, so cut the canes down to two or three buds. During the summer, pull leaves away from the grapes so that sunlight reaches them. Don't let the canopy get too overgrown, but don't thin it excessively. The leaves work hard all summer to convert sunlight into sugars that can be stored for the next year's crop.

Kiwis

Kiwis are striking plants that have beautiful spring growth, cream-colored flowers, and interesting round or heart-shaped leaves, depending on the species. The name comes from New Zealand, but the fruit originated in China and is also known as Chinese gooseberry. Kiwis are high in vitamins and minerals, and they can be stored for up to six months at 32° F.

Most kiwis need to be planted with both male and female vines, and pruned in summer and winter. It can take up to seven years for the vine to fruit, but if you love kiwi as

Gitanjali enjoys raspberries that she picks right off the plant.

much as I do, it's worth the effort to establish this highly productive and ornamental vine. I grow 'Hayward' kiwis in my garden and after two years the vines have grown dramatically. While I am waiting for my vines to fruit, I harvest kiwis from older vines in other people's yards.

My mother's Michigan home has some unsightly white aluminum columns at the entrance. I suggested she plant a female arctic kiwi 'Ananasnaya' and a male kiwi 'Kolomitka'. She planted them, patiently waited seven years, and then was rewarded with lots of grape-sized and intensely sweet fruit. But even before she had a harvest, she was able to enjoy the pretty foliage and the sight of a local robin family nesting in the vines. It was a small and slow solution that has yielded many rewards.

Cane Fruit

Raspberries and blackberries produce with fruiting canes, or floricanes. These are all from the genus *Rubus*, but there are many species, varieties, and hybrid types. Many parts of North America have their own native berries, such as thimbleberries (*R. parviflorus*) in the Northwest and black raspberries (*R. occidentalis*) in the Northeast. Often these native berries have lighter yields and smaller fruit than cultivated types, but the taste of each one is so different and individual that they are worth planting, especially in the wilder parts of zone 4 and the edges of zone 5.

For some gardeners, planting berries is like playing with horticultural fire because the canes can escape their boundaries if left

Christopher's Garden: WILD BERRIES

For many years, before my neighbors started building condos on the vacant lot next door, people from the wider neighborhood would come and eat the wild Himalayan blackberries that grew there. It was a sad day for the local berry-pickers when the bulldozers finished clearing out this free source of delicious fruit. However, the resilient blackberries popped up on my side of the fence, and I encouraged them to climb up the shady side.

Now that the condo construction is completed, the blackberries have rebounded and are coming back and producing well on both sides. This is another example of using edges and valuing the marginal. Crops like wild berries are important in permaculture because they require minimal management but give maximum results. However you must be diligent in preventing such vigorous plants from intruding from the wild into your cultivated garden. For something as rambunctious as the Himalayan blackberry, a machete may be the best tool. Also keep an eye out for shoots touching the ground, because they will tip layer.

unmanaged. Raspberries and blackberries are pioneer species, meaning that if there was a forest fire, they would be the first crop to regrow afterward. If you don't want them to spread, you must really be consistent with your pruning. On the other hand, they work very well in a food forest because they fill in gaps, some can tolerate shade, they reproduce quickly, and they can easily be moved from one part of the garden to another.

Blackberries

Blackberry options include boysenberry, marionberry, ollalieberry and thornless blackberry. My current favorite blackberry is 'Apache', an erect thornless variety. When I first saw this plant in a friend's garden, I was struck by its height and the fact that it produced huge fruits without requiring a trellis. I dug up a little sucker from the parent plant and now, only a year later, I've harvested dozens of giant fruits from my canes, and friends are placing orders for their own cuttings. Because blackberries are thorny and vigorous, they are best kept to the outer zones of the garden, and thornless varieties can be moved to a more prominent place in inner zones.

Raspberries

Raspberries are not just red but they may be golden, black, and even purple. I like them best fresh, on top of hot cereal or pancakes, or cooked in a desert like a cobbler. If you are lucky enough to have large yields, you can make raspberry jam, juice, or wine. The leaves can even be dried and made into a mineral-rich tea. Another permaculture advantage of raspberries is that they make a lot of new plants, which can be propagated or cut back many times a year. You can leave the clipped canes as mulch directly around the plants, or add them to the compost.

Blackberries are normally part of the shrub layer, but they can also be trained up and over a bamboo trellis as I have done in my garden with these 'Black Satin Thornless.'

Fruiting Ground Covers

The ground cover layer is a living mulch that helps to keep the soil moist, suppresses weeds, and provides an attractive carpet for the garden. It's even better when your ground cover plants are loaded with fresh fruit. Strawberries, of course, are the fruiting ground cover plant that does well almost everywhere, and it's a permaculture staple.

Strawberries

In general, strawberries are short-lived perennials that will last a few years. The plants send out runners that root themselves to form new plants, so you don't need to prune strawberries, but you can renovate the strawberry beds periodically by pulling out the unproductive older plants and allowing the new plants room to spread.

Strawberries are an ideal plant for edging beds and pathways. They will send out runners to form a mass of new plants and provide fruit for many months.

Mulching with straw is the traditional method of keeping strawberries clear of dirt. If you find that your strawberries are plagued by lots of pests, try moving them to a different area, or even growing them in containers. 'Seascape' is one of my favorite strawberry varieties. It's an everbearing type that can start producing really large berries in the spring and continue all the way until Halloween. It has a lot of disease resistance, doesn't send out too many runners, and does well in most climates across the country.

Alpine strawberries (*Fragaria vesca*) are very small and flavorful, and they can grow in partial shade. You can also start them from seed, and they produce in the first year, so they're a good plant for children to grow. I like the red varieties, like 'Mignonette' because of their intense flavor and productivity, but alpine strawberries also come in yellow and white varieties like 'Pineapple' and 'White Delight'.

Other ground covers

On the West Coast, uva ursi or kinnikinnick (*Arctostaphylos uva-ursi*), a creeping manzanita relative, is really a good choice because it's native, drought tolerant, medicinal, and has pretty red berries and red-brown stems with green shiny leaves. Wintergreen (*Gaultheria* species) is native to the East Coast, but it can be widely grown and makes a good-tasting berry. The leaves can also be dried and made into teas. 'Emerald Carpet' raspberry (*Rubus pentalobus*) is a thornless evergreen ground cover that produces yellow fruits. In the east, low-bush cranberry and low-bush blueberry can be grown if the soil is peaty and acidic enough.

Perennial Vegetables

Perennials are a good example of how permaculture is an ecologically based gardening system that is modeled on natural ecosystems. You need only plant them once and they come back year after year. Once established, they don't disturb the soil, and they develop large root systems that pull up nutrients and water from deep down. Perennials are often the first plants to start growing in the spring, which makes them especially valuable as we emerge from the hungry months of winter.

Herbaceous perennials are those that die back in the winter and then regrow in spring. Asparagus and rhubarb are the best-known herbaceous perennial vegetables and are a welcome treat after a long winter. An established asparagus bed that is five years old or older can be harvested for up to two months a season. Rhubarb is considered a spring tonic and while the stems are the only edible part of the plant, the leaves make a lot of mulch or compost when you discard them.

Comfrey is the superstar of the herbaceous perennials as it grows quickly and the leaves can be harvested for making chicken food, mulch, compost, and liquid fertilizer. It is easily propagated by digging up and dividing.

You must plant perennials with some care as they will be in the ground for many years. Amend the planting hole with compost, and continue each fall to spread a layer of partially decomposed organic matter or sheet mulch around the plant. The organic matter will work its way into the soil during the winter and help provide extra nutrition for the plant as it prepares for spring growth. In some situations, I also add finished compost in the early spring before the active spring growth has started.

After the first flush of spring growth, most perennials continue to grow through the summer and flower in the fall. During the summer months, keep them well-watered and mulched to keep in the moisture and keep the weeds down around the

Comfrey is a much-valued plant in permaculture gardens. The flowers attract pollinating insects like bees and the leaves can be cut and left on the ground as mulch.

PERENNIAL EDIBLES BED

The variety of perennial edibles means that you can harvest fresh food for many months of the year, for many years in a row. This bed of very different perennials creates a polyculture that helps to enrich the soil and keep pests and diseases at bay.

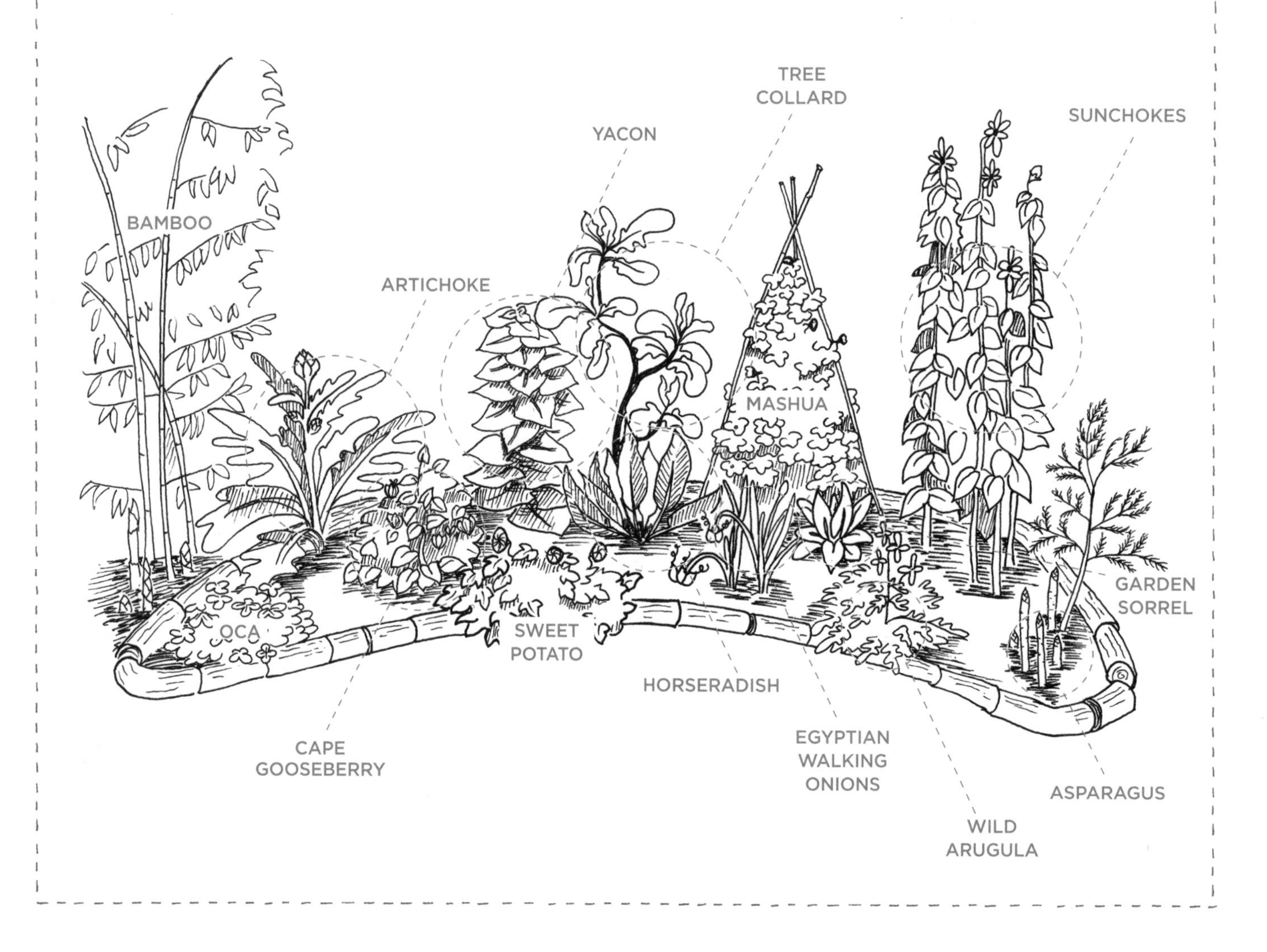

plants. If you are saving seed for propagating more plants or for sharing, allow the plants to go to seed in the fall, then collect the seed and cut back the plant. This works well for plants like asparagus, artichoke, garden sorrel, purple coneflowers, and rhubarb. A few select plants, like tree collards, will not make seed and in winter-freezing climates you will need to take cuttings inside in the fall and keep the plants going indoors until the following spring.

Divisions

Division is a very simple method of propagating new plants from the roots and crowns of existing ones. This method works for herbaceous perennials such as artichoke, asparagus, rhubarb, sunchokes, and yacon. Early spring or early fall is usually the best time to make divisions. Simply dig up the existing plant and cut back the foliage. Break or cut the crown into two or more pieces, each with growing shoots and roots. For large plants, you may need to use two spading forks acting as levers against each other to divide the root mass. Replant the new divisions in fresh planting holes or in containers of potting soil to put in the greenhouse.

To divide comfrey, dig up the parent plant and cut 2-inch pieces of the root. Place each root into a 4-inch pot filled with potting soil, or plant directly in the garden.

Asparagus and artichokes

These perennials serve as part of the shrub layer in the food forest. Some people don't much care for their somewhat grassy taste, but those who love them can never get enough during the peak early spring season. And both make striking ornamental plants.

Asparagus is one of the only cultivated vegetables that grows wild along roadsides and railroad tracks in a large part of the country. I have given up a permanent back section of zone 4 to an asparagus bed. The spears emerge in early spring when not much else is available, and I can make an entire meal of asparagus roasted with a little olive oil, salt, and pepper. My older daughter loves to eat these little treelike stalks when they're lightly steamed.

Asparagus plants are either male or female. Some gardeners find the male plants to be better producers, but I prefer female varieties because they make an excellent hedge with red berries that are good bird habitat. 'Mary Washington' asparagus is one of my favorites. It's an heirloom variety that can be grown from seed, and it has a lot of disease resistance.

Rhubarb

Rhubarb is a cool-season perennial that can be harvested over a long period. It can't be grown in climates that are too warm because it needs a period of cold to trigger growth. The plant grows from crowns, much like artichokes, and you need to give it plenty of well-rotted manure and compost when planting. The stalks can be harvested from spring through fall, and stewed or baked in pies and cobblers. The leaves, however, are toxic and should be put in the compost.

Perennial greens

Perennial greens are great for any beginner to grow because they need little care yet provide an abundant crop for many months. Their deep roots store energy so the plants are ready to emerge as soon as the snow melts. Some perennial greens, like dandelions, may already be growing in your garden. Though thought of as a weed by many, especially in the lawn, dandelion greens are

Cuttings

Stem or tip cuttings are the easiest way to propagate many plants quickly. Examples of softwood cuttings are perennials like tree collards and salvias, or annuals like tomatoes. Examples of semi-hardwood cuttings are passionfruit and kiwi. Examples of hardwood cuttings are grapes and mulberries. Cuttings are best taken when parent plants are actively growing in the vegetative or post-flowering phase. Use sharp hand pruners. Take cuttings with at least three buds. Remove most leaves, cutting back any remaining leaves, and don't cut the tip. Bury up to two buds in perlite and regularly water. Don't forget to label.

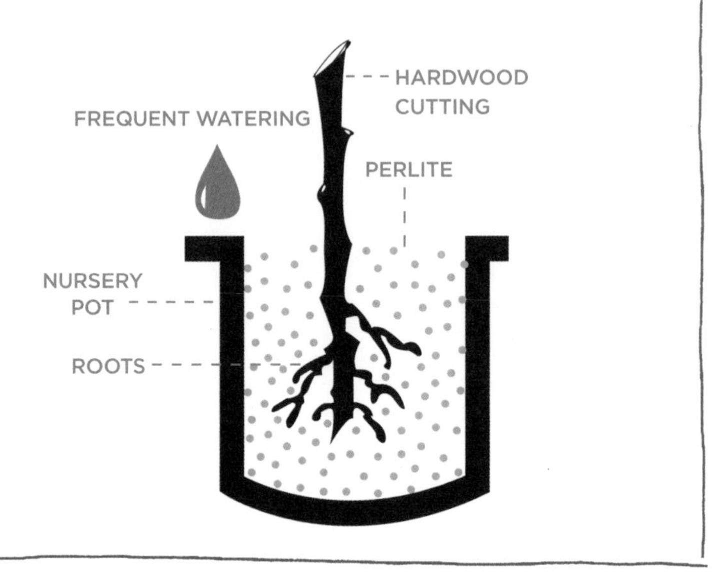

highly nutritious and can be included in the permaculture garden.

Purple tree collards are my favorite greens. These unusual plants are part of the Brassica family, just like cabbage and kale (tree collards are sometimes listed in seed catalogs as giant kale). Unlike their herb layer cousins, they can grow up to 12 feet tall and they live for many years. As part of the canopy or small tree level of the food forest, they tower above most other edibles. But because they don't cast too much shade, they create their own mini food forest that allows you to plant many things beneath them.

Tree collards have green leaves with purple veins, and the taste is more delicate and sweet than regular collards. I eat them twelve months out of the year and have intentionally overplanted them in my garden to help feed my chickens and ducks—the fowl get the larger, tougher leaves, and I save the more tender leaves for the family. As you'll frequently harvest the leaves, other crops can be trained to climb up the stalks, such as cucumbers, beans, or mashua. Lower-growing edibles like tomatoes or broccoli can be grown in the tree collard understory.

You can easily grow tree collards from cuttings. Take the cuttings in late fall and start them in a greenhouse, or, in mild climates, directly outdoors in garden beds.

Purple tree collards are towering plants that I grow extensively in my garden. On the left is mashua, a perennial root crop that originated in the Andes.

Fiddleheads

These delicacies are simply the emerging fronds of certain ferns. Fiddleheads make

an ideal herb layer in the food forest, especially in damp, rich soil under the tree canopy. Several species of fern are edible, but the most common one grown for fiddleheads is the ostrich fern (*Matteuccia struthiopteris*). They can be widely grown, tolerating both cold winters and hot summers, as long as they have partial shade and plenty of compost and mulch. They also need a fair amount of room, so they are a good plant for zone 4. Harvest a few of the young fiddleheads from each plant in the spring, leaving other fronds to grow fully. The fiddleheads can be boiled or steamed and eaten right away, or blanched and then frozen.

ROOTS AND TUBERS

Roots and tubers store well either in the ground or in a root cellar, basement, or garage, providing something tasty for soups and stews when not much else is available during winter. Most people are familiar with carrots and potatoes, but there are other nutritious and delicious root crops that are easy to grow. These storage crops range from ground covers to vines, shrubs, or giant flowers, so they work to fill many levels and niches in the food forest.

Andean root vegetables

Three of my favorite root crops are a South American version of the three sisters: mashua, oca, and yacon. Mashua (*Tropaeolum tuberosum*) is a beautiful, nasturtium-like vine that produces peppery edible tubers, spicy leaves, and mildly piquant flowers, so it's a food plant that can be used in many ways. Oca (*Oxalis tuberosa*) is a perennial ground cover that makes an attractive border. In Bolivia and Peru this Andean root crop is second in popularity only to the more famous potato. The tubers may be pink, orange, red, purple, white, or

Miner's lettuce is a wild plant that is often considered a weed. But it makes a nutritious addition to salads or can be lightly steamed. Find out what wild plants are edible in your area by consulting local wild-food experts, Master Gardeners, or regional botanical gardens.

Edible Weeds

Dandelions aren't the only edible plants that some consider weeds. There are many other plants that some disdain, but which almost always produce young tender leaves that can be enjoyed fresh or cooked. Follow the permaculture principle of observing and interacting, letting some areas of your garden go wild and seeing what you can harvest. Also check with your local Master Gardeners and regional botanical gardens for guides to wild edibles in your area.

cheese mallow (*Malva sylvestris*)
chickweed (*Stellaria media*)
chicory (*Cichorium intybus*)
curly and yellow dock (*Rumex crispus*)
dandelions (*Taraxacum officinale*)
epazote (*Dysphania ambrosioides*)
milk thistle (*Silybum marianum*)
miner's lettuce (*Claytonia perfoliata*)
mint (*Mentha* species)
pigweed (*Amaranthus* species)
pokeweed (*Phytolacca americana*) (must be blanched and cooked)
plantain (*Plantago* species)
purslane (*Portulaca oleracea*)
sheep sorrel (*Rumex acetosella*)
sow thistle (*Sonchus* species)
stinging nettle (*Urtica dioica*) (must be cooked)
wild lettuce (*Lactuca virosa*)
wild mustard (*Sinapsis arvensis*)

yellow, depending on the variety. You can eat the young leaves and stems as greens, or boil, bake, or fry the tubers. Oca needs a frost-free climate and a long growing season. Yacon (*Smallanthus sonchifolius,* also called Bolivian sunroot or Peruvian ground apple) is a herbaceous perennial, with dramatic, arrow-shaped leaves and regal, quarter-sized yellow-orange sunflowers in late fall. The tubers are very sweet, crunchy, and juicy. You can peel them and eat them raw, or stir fry them for five minutes.

Sunchokes

Another perennial root plant is Jerusalem artichoke or sunchoke (*Helianthus tuberosus*). These sunflowers are easy to grow and require a fair amount of space, as they tend to spread and can reach up to 10 feet in height. They are drought-tolerant plants that require little care. The tubers are sweeter if harvested after the first light frost. Cook them much like potatoes, either peeled or with the skin on, by baking, boiling, roasting, stir frying, or pickling.

Horseradish

Horseradish root is mainly used as a seasoning, but like sunchokes, horseradish can get a little too vigorous and take over its space. To avoid having much more horseradish than you can possibly eat or share, grow the plants in a large tub, such as a barrel planter. If you have the room, horseradish can tolerate partial shade and so it makes a good understory plant for the food forest. The one-year-old plants have the best flavor, so you may want to plant new roots each year, in fall. Use the larger roots for eating and store the smaller roots (about the thickness of a pencil) for replanting the following spring.

This Andean version of a three sisters planting includes mashua (BOTTOM LEFT), *oca* (TOP RIGHT), *and yacon* (TOP LEFT). *All produce edible tubers that can be cooked and enjoyed in a variety of ways.*

BOTTOM RIGHT
Sunchokes, also called Jerusalem artichokes, are harvested when the leaves of the plant have died back.

YACO

Mushrooms

We don't usually think of mushrooms as a crop, but in permaculture they occupy their own layer in the food forest. Not only can you enjoy mushrooms as food, but they form an important part of the soil food web, adding nutrients to the soil. The mycelium provides food for various creatures, including bees and worms. Some growers use mushrooms as a filter to clean contaminants from the soil. In fact, some growers use king stropharia ('Garden Giant') mushrooms as a bioremediator to clean contaminated water. Greywater and pasture runoff can be filtered through woodchips infiltrated with king stropharia mycelium. As the water passes through this filter system, bacteria and nitrogen are removed and clean water is produced.

Once established, mushrooms are work-free. They flourish in a variety of growing mediums and climates, so they can be grown in almost any region. Many mushroom varieties, like oyster, shiitake, and lion's mane, can be grown on logs tucked into shady, damp places, outdoors or in a greenhouse or shed. A good permaculture planting might be tomatoes as an overstory layer, kale and chard as a low tree equivalent, and mushrooms in the dark damp understory of the vegetable bed.

Mushrooms can be eaten, or dried and made into a tea, but it's obviously important to know what you are harvesting. Never eat a mushroom that has appeared in the food forest unless you are absolutely certain what it is. Many edible mushrooms resemble species that can cause serious illness. It's well worth the time to attend a local workshop on identifying and growing mushrooms so that you know the common types in your area.

Hardwood logs like oak provide a good growing medium for mushrooms. You can even make a fence do double-duty as a mushroom farm.

Making Mushroom Logs

One of the best ways to grow your own mushrooms is from wooden plugs that have been colonized with mushroom spawn. These plugs are available from commercial suppliers and are inserted into hardwood logs where they inoculate the host wood. The mycelium grows through the log, and eventually it begins to produce mushrooms. If you don't have your own hardwood logs (such as alder, cottonwood, oak, elm, maple, or poplar) you can usually get cut logs and stumps from landscape suppliers or arborists. Do not use aromatic logs like cedar or pine, nor thin-barked woods such as birch. Logs should have been cut at least three weeks before you insert the plugs, but well-aged logs are not good candidates.

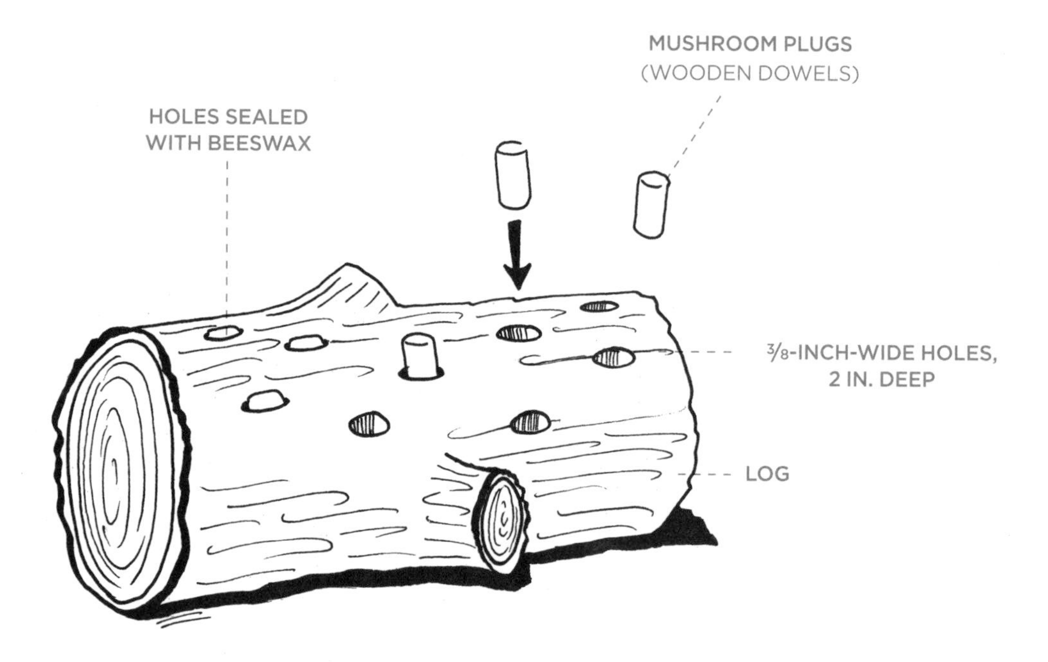

Materials

mushroom plugs
6- to 12-inch logs or stumps
beeswax
old saucepan
metal cup
gloves

Instructions

1. Lay mulch to a depth of 6 inches.
2. Lay the logs on the bed so that they are well-settled in the compost. In damp climates, you can set them upright as they will not dry out.
3. Heat beeswax in the saucepan on a stove at a low simmer.
4. With a $^3/_8$-inch drill bit, drill holes 2 inches deep, about twenty holes per log. Rinse the holes with sterile water.
5. Hammer a wooden plug into each hole.
6. Pour hot beeswax into the metal cup; then pour a small amount onto each plug end to seal it.
7. Leave the logs in place; it may take several years to produce mushrooms.

Plant young seedlings outdoors when the weather has warmed. Protect them from frost with covers like these glass cloches, and mulch the ground heavily with straw.

Annual Vegetables

In permaculture we rely on trees, shrubs, and perennials to form the basic framework of our food forest. However, edible gardeners also have many annual favorites, and there is plenty of room for annuals in the permaculture garden. You can build raised beds for annual crops, grow them in containers, or mix them into your tree guilds. In particular, leafy annuals serve as part of the herb layer, and if left to seed, bring a large array of beneficial insects. Annuals grow quickly and are much faster at adapting to changing conditions, which is another reason to choose regionally adapted annual seeds.

Annuals are typically described as cool-season and warm-season, although in some areas, cool-season crops can be grown throughout the year. Planting dates typically depend on when your first and last frosts occur. Most areas across the country are frost-free from June to September, but in some areas, planting as late as July may be necessary to avoid frosts. In other areas, a few hardy crops like radishes, onions, lettuce, and spinach can be planted as early as April. I have a friend in the Midwest who picks Brussels sprouts out of snowdrifts for Thanksgiving. The more you talk to other gardeners and experiment, the more knowledge you'll have about how and when to grow successful annual crops.

Dark leafy greens (beets, chard, mustard, turnip, and spinach)

When I think of hardy plants in the herb layer that are easy to grow and pack a nutritional punch, dark leafy greens come to mind. They're filled with vitamins and minerals (such as calcium, iron, and zinc) and come in many sizes, shapes, and colors. You

ANNUAL EDIBLES BED

By midsummer, an annual bed can contain a wide mix of crops, including tall growers like tomatoes and beans that must be trellised, leafy greens, and root vegetables. Members of the fungi layer may colonize the logs that surround the bed.

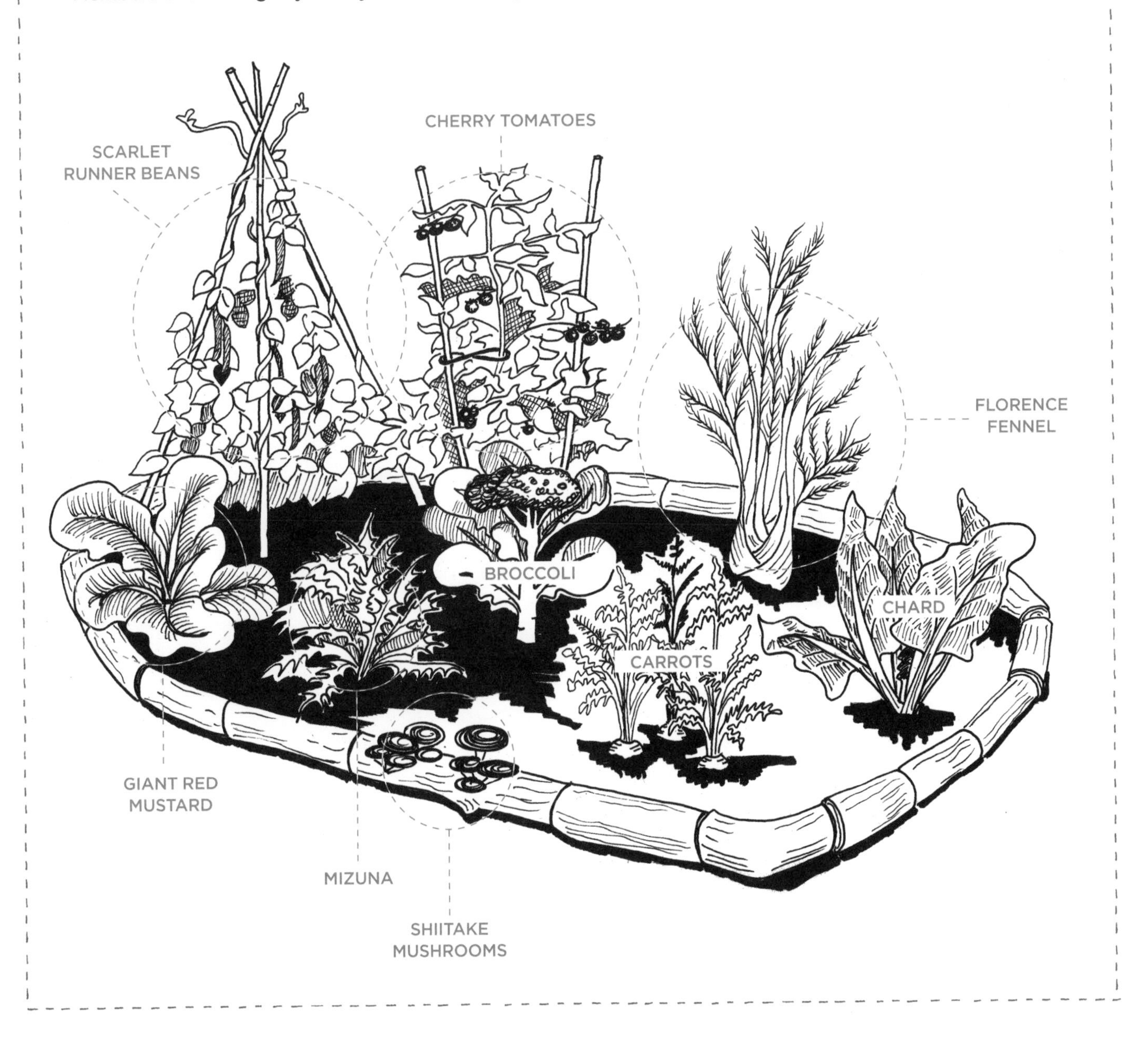

get a speedy harvest, which can often be extended as you can pick repeatedly over many months.

Chard is one of my favorite dark leafy greens for a variety of reasons. It is related to spinach but is hugely productive and less fussy. Once established, it requires very little maintenance, it can be picked again and again for months, and both the leaves and stalk can be used in a variety of recipes. My favorite variety is 'Five Color Silverbeet'.

Beets and turnips are grown for their edible roots, but the greens can also be harvested. If you grow these crops from seed, you can eat the seedling thinnings fresh. The more mature greens are highly nutritious and can be cooked much like kale or chard. Beets can be sown through the growing season from spring into winter and you can harvest greens just a month after sowing from seed.

Spinach is a cool-season crop, although you can get bolt-resistant varieties for growing into the summer months. Or plant it as an understory crop around tall summer crops like beans or tomatoes. Malabar spinach (*Basella alba*) is a good spinach alternative in hot areas. It has a vining habit, and its red stems, white flowers, and purple fruit are highly ornamental. Orach (*Atriplex hortensis*) is sometimes called mountain spinach as it originated in the Alps. It's another warm-season salad green similar to spinach. Pick the leaves before they reach 18 inches in height. The seed heads are very ornamental and attract beneficial insects.

Salad greens (corn salad, lettuce, mesclun, mizuna, and spinach)

Salad greens are highly productive as cut-and-come-again plants. I love snipping greens for my salad on a daily basis, and

LEFT
This polyculture planting of annuals combines peas, cabbage, and greens along with strawberries. This diversity helps keep pests in check.

RIGHT
Many crawling and flying insects like to nibble on the leaves of greens and plants in the brassica family. Be willing to accept a certain amount of damage and let natural checks and balances keep the pests to minimal numbers.

Cool- and Warm-Season Crops

COOL SEASON: Arugula (rocket), beet*, broccoli*, cauliflower*, cabbage*, celeriac, Chinese cabbage, carrots*, chard*, chicory (endive, frisee, and radicchio), collards*, fava bean, Florence fennel, garlic, shallots, kale, leek, lettuce, mizuna, mustard*, spinach, mangle*, mache, mesclun (salad mix), onion*, pak choi, parsnip, pea, potato*, radish, rutabaga, salsify (oyster plant), tatsoi

WARM SEASON: pole and bush bean, runner bean, corn, cucumber, eggplant, melon, okra, sweet and hot pepper, summer squash, winter squash, tomato, tomatillo, turnip*, watermelon

*In some climates, these can be grown in summer or winter, depending on the variety and local growing conditions.

even in extreme climates, salad greens can be grown outside for many months and in cold frames, greenhouses, or on windowsills in the winter. In permaculture, salad greens, like dark leafy greens, are part of the herb layer, almost to the low level of a ground cover. Salad greens are a good planting to use around the edges of beds and under taller plants. Putting them on the edges also allows you to harvest them easily.

For continuous harvest, salad greens need to be planted in succession. I once worked on a farm where we planted 'Black Seeded Simpson' lettuce every two weeks. It's a giant green lettuce with lots of pretty light-green, wavy leaves that have plenty of flavor and no bitterness. It stores well, so you can pull the whole plant out of the ground and make enough salad for almost a week, or pick the outer leaves as needed. It is also heat tolerant, so can be grown in warmer months.

Another favorite is 'Speckled Black Trout' lettuce, a green romaine type with red-splashed leaves. Its buttery flavor is the perfect contrast to more pungent salad ingredients like spicy radishes or peppery arugula. Two varieties of lettuce developed at the University of Hawaii are excellent for beginners: 'Manoa' is easiest to grow, and 'Anuenue', a butterhead-iceberg cross, comes in a rainbow of colors, has a crispy texture, and tolerates warm weather.

For an Asian salad mix, try mizuna, a cut-and-come-again salad green with fine, feathery leaves and white stems. In Japanese mizuna means "water/juicy vegetable" and it is often pickled, but with its mild and tangy taste it is also excellent as a fresh salad green. I've also used it in stuffing for chicken and ravioli.

In colder climates, you can manage to grow salad greens year-round by using plastic hoop houses or cold frames. 'Winter Density', 'Arctic King', and 'Red Salad Bowl' lettuce are all good options for winter plantings. Floating row covers will help protect salad beds from early or late frosts.

LEFT
'Rainbow' chard is ornamental and tasty, with crumpled dark green leaves and stalks in hues of red, orange, purple, yellow, and white.

ABOVE
My nephew Christopher Wallace with a crop of mustard greens. These can be eaten and left to produce flowers to bring beneficial insects.

Asian greens (bok choi, Chinese cabbage, Chinese red mustard, pak choi, tatsoi) Many people are familiar with stir-fry staples like bok choi and Chinese cabbage, but the Asian greens category includes many unique varieties that are worth growing. Not only do they add to the herb layer of your garden, but seed saving from these plants results in more beneficial insects. Asian greens are fast to produce, packed with nutrition, and taste great with a little sesame oil, ginger, and garlic. You can find many seed mixes of Asian greens through mail-order seed suppliers. I buy seeds from Kitazowa Seeds in Oakland; they have a diverse array of Asian vegetables to order online (kitazowaseeds.com).

My favorite Asian green is 'Red Giant' mustard, with striking deep purple and green leaves, which are peppery and sharp when eaten raw, yet only mildly spicy when cooked. This broad taste continuum is one of the things I love about this plant, in addition to it being one of the most nutrient-dense vegetables around. It reseeds easily and will not cross-pollinate with the more familiar brassicas like broccoli, so it's a great plant to add to your self-seeding permaculture garden.

The spicy Korean condiment *kimchee* is derived from the Chinese cabbage, 'Aichi' (also called Napa cabbage), which prefers cool weather. It produces large barrel-shaped heads that are more tender and mild than European cabbage. The young leaves can also be used in salads, stir fries, and soups.

A loose-head type of Chinese cabbage with a mild and delicate flavor is 'Beka Santoh'. The first time I grew it, I was sure it was a loose-leaf lettuce until it produced some recognizable yellow brassica flowers. It is incredibly fast to produce edible leaves (only twenty-five days from seed).

Biennials (endive, fennel, frisee, and radicchio)

Increase your culinary skills and entice your taste buds with this palate of western European traditional vegetables. These biennials also form part of the herb layer, and take two

LEFT
Mulch young lettuce plants with leaves or straw. Watch for slugs and snails, which are the main threat to your salad greens.

RIGHT
Tatsoi and shallots in a permaculture bed.

TATSOI

years to complete their growth cycle. If left to flower and seed, they can become habitat for butterflies and other beneficial insects.

Endive, frisee (curly endive), and radicchio are sometimes mislabeled on seed packets, but they are all slightly bitter-tasting greens that make wonderful salad ingredients. You can also roast endive and radicchio in the oven with a little olive oil and balsamic vinegar, or add them to other roasted vegetable dishes. Radicchio typically has a tight head, but some varieties are more loose-leafed. These vegetables can all be grown from seed, but they do need consistent moisture during the summer to avoid bolting.

Florence fennel is a showy vegetable that produces a white bulb above the ground, and edible feathery green leaves that look like dill. It has a licorice-like flavor that mellows and sweetens when roasted, but you can also add the sliced bulb to stir-fries and salads.

SOLANUMS AND OTHER HOT-SUMMER CROPS

Tomatoes

It is for the love of tomatoes that many people become home gardeners. In the permaculture garden, these and other members of the solanum family are part of the herb or

Florence fennel needs plenty of water through the growing season, but keep mulch away from the bulbs as it can lead to root rot.

LEFT

Sow seeds of salad and Asian greens successively through the growing season so that you always have some cut-and-come-again leaves for the salad or stir fry.

the shrub layer, depending on their height. Tomatoes were first domesticated by Mayan farmers thousands of years ago, and have been grown in the U.S. for many years by indigenous peoples in the Southwest. Today, it's the most popular edible grown in this country. I heard that someone in San Francisco grows heirloom Russian tomatoes on the fifty-seventh story of the Transamerica building. Now that's some vertical growing!

There are a variety of tomato types and how you plan on using them should guide which types you grow. In seed catalogs, tomatoes may be categorized by maturity dates (early, midseason, or late season), size, color, shape, culinary use (slicing, drying, or paste), and whether they are hybrid or heirloom plants. They are often divided into determinate (bush) and indeterminate (vining), which affects where you will plant them and whether they'll need support. Each growing season is different, so try several types each year; it's unlikely you'll ever settle on just one type of tomato.

The one requirement tomatoes have is sun—they really love lots of sunshine, so find the sunniest spots in the garden, away from any tree canopy. In my cool summer coastal climate, Russian and Eastern European tomatoes—mostly short-season types—do well. 'Siberian' is a determinate variety that needs little staking and produces all its tasty small fruit in a short time. 'Paul Robeson' is a dark and flavorful large indeterminate tomato. (I also like the fact that this Russian-developed tomato was named after the American singer and activist.) 'San Marzano' is a nice paste tomato, good for canning, drying, and oven roasting. 'Roma' is another variety good for sauces and salsas. 'Early Girl' is an excellent choice as a slicing tomato for salads and burgers. 'Brandywine' does best in the eastern U.S. and wins many flavor contests. 'Cherokee Purple' is a comparable favorite for the West Coast. In climates with hot summer nights, any beefsteak type will do well. 'Peacevine' is a fine-tasting and prolific cherry tomato, and one plant is all you need. Try drying cherry tomatoes for a week in the sun between two window screens.

Tomatillos are the essential ingredient in salsa verde, and they grow much like bush tomatoes. The fruit is surrounded by a papery husk and is somewhat tart even when ripe. Although the fruits are not large, they do require a long summer to ripen.

Tomatoes usually need some support. You can use cages, bamboo trellises, or a string trellis, as shown here. This kind of trellis is easy to build—all you need is a frame and some garden twine.

OPPOSITE
There are tomato varieties for almost every garden. My garden has cool summers so I grow smaller varieties that have time to ripen.

Brewing Liquid Fertilizer

Tomatoes are heavy feeders and can benefit from added nutrition during the growing season. Rather than bringing in commercial fertilizer, you can make your own on a weekly or monthly basis by taking certain nutrient-rich plants and brewing them into a water-soluble plant food. Making your own liquid fertilizers is another way to use and value renewable resources. Apply the fertilizers regularly to heavy feeders during the spring and summer, tapering off in the fall so that all new growth can harden off before winter.

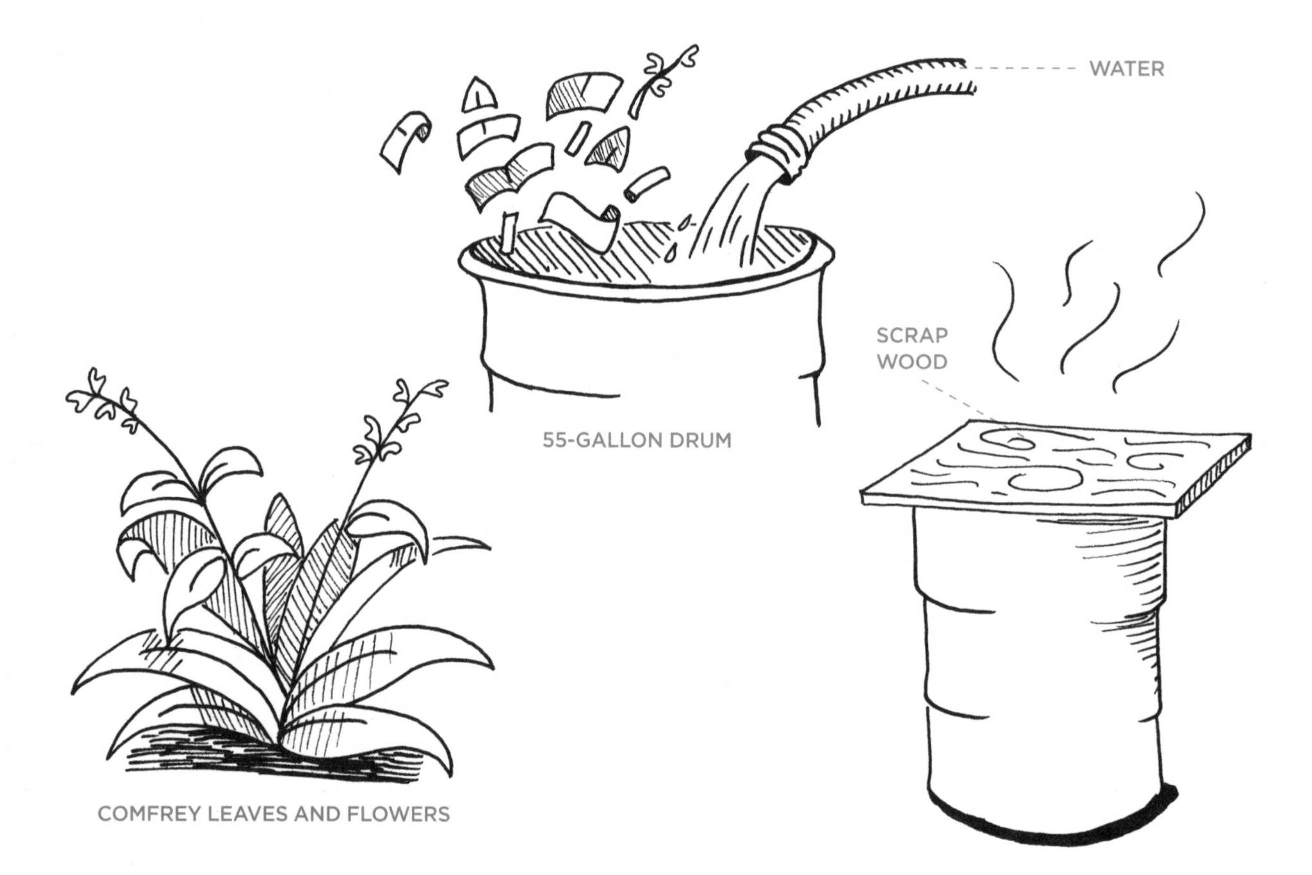

Materials

55-gallon drum with lid (food grade)
5-gallon bucket (food grade)
garden hose (or water from rain barrel)
sharp knife or hand pruners
mineral-rich plants such as chicory, comfrey, dock, dandelion, horsetail, nettles, or yarrow

Instructions

1. Harvest greens from garden. Cut off the leaves rather than digging up the entire plant.
2. Put the greens into the container, a little less than half full (no more than 20 gallons).
3. Use your feet or a tool to compact the greens.
4. Fill with water, and cover securely to keep out animals. The brew will start to smell strongly in a week or even sooner in warm weather.
5. Scoop out a 5-gallon bucket's worth of the brew and leave the decaying plants in the large drum. You can keep topping off with new water for a week until the smell becomes too much. Put the remaining material in the compost.
6. Filter the liquid fertilizer through a paint filter and put it into a pump sprayer.
7. Foliar plant sprays are better absorbed by the leaves if they are already moist. Spray plants first with plain water. When applying the liquid fertilizer, try to cover both sides of the leaves. Liquid fertilizer does not store well, so use immediately.

Christopher's Garden: BUTTERFLIES

My daughters and I find the tiny caterpillars of anise swallowtail butterflies when we harvest Florence fennel and bronze fennel. We carefully pluck them off the leaves and bring them into the house. We keep them in mason jars (with holes punched in the lids for ventilation) and continue to feed them fennel, until they turn into chrysalises. Once they emerge as butterflies, we release them back into the garden.

Swallowtail larvae.

OPPOSITE
Two cucurbits from my garden: a 'Tromboncino' zucchini harvested by Gitanjali and a buttercup type growing on a bamboo trellis.

Peppers and eggplant
More sun-loving South American natives, peppers and eggplants need a long, warm growing season. In cooler climates, you can grow them in the greenhouse, or grow them indoors until the temperatures are warm enough to set the young plants out in the garden. If you are saving pepper seeds, keep your hot and sweet peppers apart to avoid cross-pollination. Peppers and eggplant will be in the ground for many months, so use the space underneath them to plant annual greens or heat-loving herbs such as basil.

Cucumber, squash, pumpkin, and zucchini
When most gardeners think of summer, they think of endless zucchini. As a kid, I remember wheeling one wheelbarrow full of zucchini after another from my 4-H garden plot to drop off on my neighbors' porches. One reason this plant family is so prolific is its versatility. These sprawling plants can be part of the ground cover layer in the food forest or you can maximize space and make them part of the vine layer by training them up trellises or over other supports.

Cucumbers are ideal for trellising, and in colder climates can even be trellised in a greenhouse. These warm-season crops need heat to sprout and develop, and many types have pollination requirements, so check on their needs before you choose your varieties. One permaculture combination is to grow cucumber on a trellis, or even a sturdy frame set at an angle, and to use the shady area below for summer lettuces, spinach, and other greens.

There are endless choices when it comes to squash and pumpkins. Summer squash include the scalloped or pattypan types, crookneck, and zucchini. These are harvested in summer when the weather is still warm and the fruit is still tender and thin-skinned. Winter squash—like acorn, banana, and spaghetti types—are grown for harvest in fall; they store well and are often used for baking. Pumpkins may be vining or bush types and they all contain edible seeds in addition to the flesh.

Squash plants call for a lot of room, because they like to spread. Fortunately, they can be grown out of a *hugelkultur* bed or in other slightly wild zones of the garden. As the large leaves spread across the ground, they retain moisture in the soil, providing their own mulch. Just keep the fruits off the ground to prevent them from rotting by placing them on a wooden board or bed of dry straw. My favorite squash varieties include

'Banana' (big and flavorful), 'Kabocha' (high-quality, small fruit), 'Buttercup' (best in the Midwest), 'Orange Hubbard' (best for short growing seasons), and 'Table Queen' (a bush acorn for small spaces).

My favorite winter squash is 'Waltham Butternut', just the right size for a meal and then some leftovers the next day. It has a rich and nutty flavor and high yield, with 3- to 6-pound fruits that can be stored for a long time.

'Black Beauty' is my favorite zucchini, loading its vines with shiny, dark green, almost black fruits. Keep picking them when they are under 8 inches long, or you end up with giant specimens only good for stuffing and baking.

Corn

If you have the space and the summer temperatures, corn is a deeply satisfying plant to grow. Just planting a few small seeds yields enormous dark green plants that provide

sweet summer treats. In the permaculture garden, corn can also provide a trellis for other climbing summer crops, such as beans.

One summer at a community garden in Detroit, I was shown a Native American corn-planting technique by gardener Gerald Hairston, whose grandmother was a Blackfoot Indian. She taught him to bury fish heads and bones on a mound, then plant corn on top. Corn is a heavy feeder, so it benefits from the extra nutrients. If you don't care to plant fish heads with your corn, feed the plants weekly with liquid fertilizer.

Corn grows in a wide range of climates, even cold climates, thanks to short-season varieties that have been developed by breeders. Your local growing conditions will help you determine which varieties will do best. My favorites are 'Painted Hill Sweet Corn' which is good for northern growers with short growing seasons and cool spring soils. 'Double Standard' is an open-pollinated sweet corn and 'Old Country Gentlemen' is also a good choice.

Okra

Okra is a heat-loving annual vegetable that just doesn't do well in my cool-summer coastal garden, so even though I've tried to grow it, I haven't had much success. In warm-summer areas, it forms large, bushy plants that can grow up to 6 feet in height. Okra is related to hibiscus so it has really ornamental flowers in a range of colors, depending on variety. There are some very interesting heirloom varieties, including some that produce deep purple pods. 'Burgundy' has red stems and pods and can be grown in a large container.

You can start okra indoors a month before the last frost to get a good head start on the season and to protect the young plants from birds, snails and slugs. Set the plants in the garden when it's good and warm, about 70° F. Give the plants plenty of water and top-dress with compost during the growing season.

The trick in getting the best okra pods is to harvest them when they are slightly immature, and not let the plant grow as large as it wants to. Size varies greatly depending on variety, but pods average 2 to 3 inches long, and they emerge just four to six days after flowering, so you have harvest regularly to keep the plants producing.

The pods can be cooked in soups, stews, and stir-fries and feature in many different types of ethnic cuisine, from Indian curries to Cajun gumbo.

ROOTS AND TUBERS

Highly nutritious plants like beets, carrots, potatoes, and sweet potatoes are part of the root layer of the food forest, growing down while most other crops grow up. These edible storage roots and tubers can be kept in pantries and root cellars for use throughout the year. I include garlic, onion, and shallots as part of this category of hard-working vegetables.

Carrots

Most young children love carrot sticks and will happily snack on them, especially when they come from the sweet tender young carrots in your own garden. 'Scarlet Nantes' are sweet and crunchy, orange-red carrots developed in France in the 1850s by the Vilmorin seed company. They can be frozen or juiced, and we also love them steamed, stir fried, or pickled. 'Dragon' carrot, bred by John Navazio of the Organic Seed Alliance, looks like the most promising red variety, with a sweet and almost spicy flavor.

Christopher's Garden: GROWING CHAYOTE

Chayote (*Sechium edule*) is a pear-shaped squash that has been cultivated in Mexico for many years. When I first planted it, I had limited success producing a crop that didn't have a fibrous texture. One day I saw my neighbor—a very old Mien (tribal Vietnamese refugee) grandmother—climbing my hawthorn tree barefoot to harvest the chayote. What I learned from her, in an animated conversation as she speaks no English, was to pick the chayote fruits more frequently when they are smaller rather than letting them grow. She taught me that this forces a higher yield from the plant, and the fruit is more succulent and never fibrous. She left with a bag of chayote and I learned a better growing technique.

After this, one of my students from Taiwan, Shen Linn, showed me how to eat the young shoots of the chayote as a stir-fried green, which she called dragon whiskers. Then a Salvadoran friend, Jose Rivas, told me that to him the best part of the plant is the massive tubers it produces, which are sweet when baked. These stories of the chayote embody the essence of permaculture, which goes beyond mere gardening. It includes sharing both knowledge and food, learning from other cultures and appreciating the entire plant.

Chayote squash.

Beets

'Detroit Dark Red' is my favorite beet variety because its color is such a dark red that the beets are almost purple-black. It's not just that I have an affinity for all things organic from Detroit, I also like the fact they get quite large so I can steam them with some butter or make beet pickles. You can also eat the beet tops, either the young thinnings, which can be put raw into salads, or the more mature leaves, which can be cooked like spinach.

Potatoes

Potatoes are perennial plants that originated in South America, but they are often grown as annuals. They are planted from so-called seed potatoes, which you must get from a reliable source to be sure they are free of viruses and other diseases. Just 2 pounds of seed potatoes can give you 50 pounds of edible tubers.

The type of potatoes you grow will depend partially on how you like to cook them. The starches in different varieties make them suitable for boiling, baking, or mashing. There are also types that are adapted to heavy, wet soils, and that have disease resistance for the many soil-borne diseases that can affect the crop.

Potatoes are typically grown in furrows amended with plenty of compost and well-rotted manure. As the plants grow, you must keep the developing tubers covered with soil or mulch.

One method for growing potatoes if you don't have space in your beds is to build a wire cage filled with leaves, compost, and well-rotted manure. You can also grow tomatoes on the outside of the cage.

Carrots need a light, crumbly soil—so dig in plenty of compost to the soil but it must be finished compost as fresh compost or manure can cause the roots to fork.

Sweet potatoes

Sweet potatoes are tropical vines with a very different growth habit than ordinary potatoes. They are actually much easier to grow—if you have the right conditions—and they suffer fewer pests and diseases. Sweet potatoes need sandy soils, plenty of sun, plenty of room, and enough moisture and nutrients—but not too much. Excessive applications of manure or other nitrogen-rich material encourages leaf growth rather than tubers.

Sweet potatoes are grown from cuttings, or slips, and unless you have well-drained soil, plant them on a mound or ridge of soil. Give them plenty of straw mulch to keep down weeds until the plants spread out. The leaves of sweet potatoes are edible, but they must be cooked. The tubers take three or four months to mature and you must harvest them before the first frost. The flavor of sweet potatoes improves if you first cure them in the sun for a day or two, then keep them in a warm, humid place for up to two weeks. After that they can be stored for many months, providing a sweet treat in the winter months.

BEANS AND PEAS

Jack of beanstalk fame isn't the only person to realize that beans are infused with magic. Both beans and peas go from humble seed to skyward vine in a matter of weeks. Some, like scarlet runner beans, can reach up to 15 feet if climbing a wall or trellis. These self-seeding annuals come in a wonderful diversity of colors and patterns, are highly productive, and form part of the vine layer.

Beans and peas are nitrogen fixers which means that bacteria living on their roots pull in free nitrogen from the air and make it available to the plants in exchange for room and board (water and sugars). This

Scarlet runner beans need a sturdy trellis to support the weight of the long bean pods. I love to see how the crimson and black dried seeds develop into these abundant vines.

exchange enriches the soil, preparing it for more demanding crops like tomatoes.

Beans

I always plant versatile scarlet runner beans, as the plants are giants in the garden, with copious bright red flowers that lead to yields of dark green beans. These beans are from the Andes and do best in cool summer areas, but can grow nearly everywhere, as my mother in Michigan has demonstrated. Hummingbirds and butterflies are drawn to the flowers, so they are excellent for attracting pollinators. Runner bean pods can be eaten when they are young and tender, either raw, cooked, or as pickles. The large immature fresh green beans can be cooked into a nice stir-fry. When the dried bean is cooked, it swells into a 2-inch bean and has a flavorful, meaty taste that is excellent for soup and burritos. Scarlet runner beans are a good recipe substitute for lima beans, which need more summer heat to grow well.

There are so many other kinds of beans to experiment with in the permaculture garden that you will never run out of choices. Dried beans are grown until the pods turn dry and brown; then they can be stored for use in soups, stews, and many other dishes. There are heirloom varieties that are well suited to all regions, including the desert Southwest. Snap beans have tender green, yellow, or purple pods that are eaten whole, steamed, or added to stews. You can also pickle them for a winter treat. Bush-type beans don't need any support and bear earlier than pole beans, which tend to be more productive but must be trellised.

Fava beans, also called broad beans, are one of the easiest and most productive vegetables to grow. These cool-season beans can be planted in fall for a spring or summer crop, or in early spring for a late summer crop. They make a good choice for *hugelkultur* beds, or in an area that has just been sheet-mulched because they germinate easily from seed and can be dug into the ground after harvest. In fact, they are often grown as a cover crop. Fava beans are subject to black aphids, which can be dislodged by spraying with water from the hose.

Peas

Peas are cool-season crops that are incredibly easy to grow from seed. Some are grown for the edible pods (snowpeas), others for the seeds inside (shelling peas). Snap peas can be eaten whole, pod and all. Plant peas in early spring as soon as the soil warms up, in late summer for a fall crop, or right through the winter months in the mildest climates. Vining types will cling to trellises or wires with curling tendrils. Bush types don't require any support, although they can cover low trellises—plant spring greens below them. Any kind of support will work for peas—bamboo poles, sticks, string, or fencing. Make sure that the plants have good air circulation, as they tend to develop powdery mildew.

After you harvest peas, dig the spent foliage into the soil or add it to the compost pile.

BRASSICAS

Most home gardeners are familiar with broccoli and cauliflower, nutrient-dense edibles that are part of the herb layer. Brassicas are biennial, and if you are growing them for seed, they produce beautiful yellow flowers that attract beneficial insects. If flying pests like white fly or cabbage loopers are preying on your brassicas, cover the plants with some floating row cover material.

While many brassicas grow best in the spring or fall, the family has great seasonal adaptability. Collards can tolerate summer heat. In cold-winter climates, kale can be harvested even when frozen. As with Brussels sprouts, the flavor of kale and mustard is actually enhanced by light frost, which seems to draw a sugary sweetness to the foretaste.

Cabbage

These cool-season crops add to the herb layer in spring, fall, and early winter. They need a rich soil, so make sure to dig in plenty of well-rotted manure or compost.

Smooth head cabbage can be stored for many months in a root cellar. Crinkled or savoy cabbage has ruffled leaves that are quite ornamental. Cabbage need plenty of moisture, so mulch them generously. Give the young plants some liquid fertilizer about a month after planting.

Fava beans need a bit of extra treatment in the kitchen because each individual bean has to be shelled twice. But I think of this as an opportunity to enjoy a slow food task.

Building a Trellis

Trellises give small gardens a lot more vertical space and help you to maximize yields in both large and small gardens. The difference in the yield of a single row of bush beans versus a row of pole beans supported on a trellis is incredible! Trellises can give a feeling of enchanted enclosure or framing to your garden as well as making it fun to feel like you are in a jungle trek or a scene from a fairy tale in your backyard. There are many types of trellises you can build, depending only on your imagination and your skill. I like to use local or onsite construction materials, such as branches or bamboo poles.

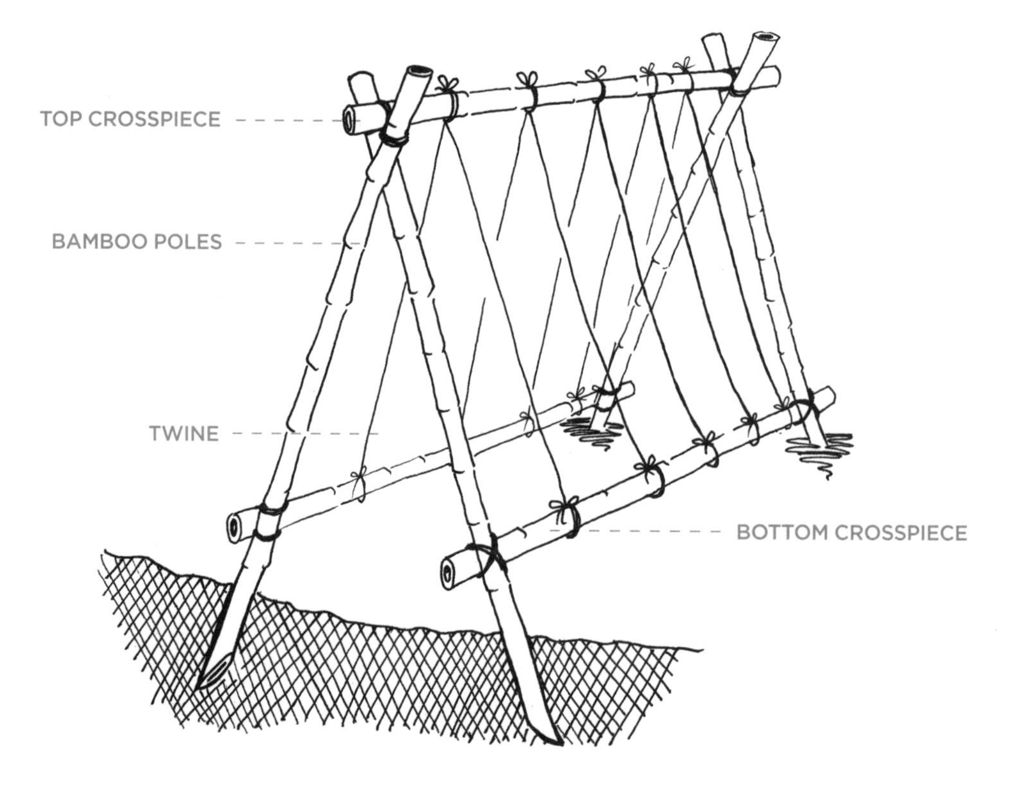

Materials

tree prunings, such as apple or plum branches or
three- to five-year-old bamboo poles, freshly harvested, branches and tops removed and soaked overnight in a borax solution (as an insect repellent)
hemp or palm twine
digging bar
loppers
pruners
saw
tape measure

Instructions

1. Cut the poles to a height of 4 to 6 feet (or to a height you can reach comfortably). Cut the bottom of the poles at a 45° angle and place the cut end into holes at each end of the trellis.
2. Make 2-foot deep holes in the ground with the digging bar. Set the poles in the holes with the angled end in the ground.
3. Cross the end poles about 6 inches from the top. Tie the poles where they intersect by crisscrossing them with twine, then looping the twine over the juncture between the two poles and pulling it tight.
4. One foot above ground level, connect all supports with a crosspiece.
5. If designed, make some X-shaped cross supports with poles (this creates triangles, which are the strongest shapes), tying them to the frame as above.
6. Finish with a top crosspiece that connects the whole trellis together, tying in the same manner as above.
7. Tie twine for plants to climb every few inches between the bottom and top crosspieces. Peas will need 2-inch spacing; beans need 6 inches. Tomatoes can be spaced every 8 inches.

Collards

Many varieties of collard are more heat-tolerant than other brassicas, but the flavor is still improved by light frosts. Use collard greens in cooking much the same way you do kale, adding the young leaves to stews, stir-fries, and soups.

Broccoli

Although the dark green variety is most common, varieties can be grown that have violet, orange, or chartreuse flower buds. 'Calabrese' is a sprouting broccoli brought to the United States by Italian immigrants in the 1880s. Last year, some of my 'Calabrese' heads actually grew larger than my own head, and fed a lot of people. 'Romanesco' is listed in seed catalogs as both a broccoli and a cauliflower. It has flavorful heads that look like whirling chartreuse green fractals. Plant them to give your garden a touch of Dr. Seuss whimsy.

Little broccoli trees are nutritious and incredibly delicious when properly cooked. Whether you are steaming, stir-frying, or sautéing your broccoli, be sure not to overcook it so that it retains a little crunch.

Kale

Kale is one of the easiest plants to grow, and it offers incredibly nutritious food during the coldest months. The only requirement it has is rich soil, so be sure to add plenty of compost and mulch the plants with straw or leaves. Like many other brassicas, kale is sweeter when it has been subject to frost. Favorites include 'Lacinato', an heirloom Italian kale from Tuscany.

TOP LEFT
Runa with a Sicilian purple cauliflower, which takes up a fair amount of room with its sprawling leaves. These are cut as mulch or compost when we harvest the head.

BOTTOM LEFT
Kale is a hardy brassica that tastes sweet after a light frost or even a snowfall. It's interplanted here with leeks. Cabbage family and onion family plants make good polyculture companions.

RIGHT
Collards can be pretty big plants—up to several feet tall and wide—so put them where they have room to grow.

Edible Herbs and Flowers

Herbs and flowers are important support players in the permaculture garden, filling in the edges with plants that range from ground covers to towering blooms. They enhance the fertility of your garden by bringing in beneficial insects and wildlife, and provide culinary treats in the form of seasonings, teas, and garnishes. I encourage beginning gardeners in particular to start with a few herbs for cooking and some reseeding flowers. Even a small deck or balcony can house these low-maintenance, high-yield plants.

Herbs are multipurpose permaculture plants, providing food and flavoring for the kitchen and attracting beneficial insects to the garden. Many, such as German chamomile, also have soothing or medicinal properties.

In the food forest, herbs form one of the main layers, blanketing the ground and keeping out weeds. Herbs are generally fast growing and can be harvested almost continuously. Many culinary herbs are perennials, providing years of yields with little inputs. Other annual herbs often reseed themselves, or you can let them flower and then save the seeds for replanting. In addition to their culinary value, herbs have many therapeutic uses, and in fact these once-marginal plants have become part of a multi-billion dollar botanical industry.

To align with the permaculture principle of using the edges and valuing the marginal, herbs can be tucked into borders and the edges of garden beds. They also make great container plants and can be grown in sunny windowsills year-round.

CULINARY HERBS

Herbs are the main determinants of the flavors we associate with regional cuisines. Many, but not all, of the best-known culinary herbs are annuals. Sweet basil and grassy parsley lend their distinctive flavors to Mediterranean dishes, while cilantro is a staple in many different Asian culinary traditions, as well as in soups, salsas, and condiments from many regions of Latin America.

Perennial herbs grow abundantly in the right conditions, and contribute to the edible landscape in many ways. For instance, if you have well-drained, sandy soil, rosemary grows into a drought-tolerant bush. Thyme is an attractive and edible ground cover that can be planted in the crevices of rock walls and between paving stones. Sage can be an essential part of mixed borders, offering fragrance and attractive variegated leaves.

HERB SPIRAL

One of the most convenient and attractive ways to design a culinary herb bed is in the form of a spiral. The spiral stacking aspect of the design allows for shade- and moisture-loving herbs such as mint to be placed at the bottom and north side of the spiral. Progressing up the midsection of the spiral are herbs with a preference for gradually hotter and drier conditions like basil, cilantro, and parsley. Topping off the spiral on the south side are heat-loving herbs like oregano, marjoram, tarragon, and thyme.

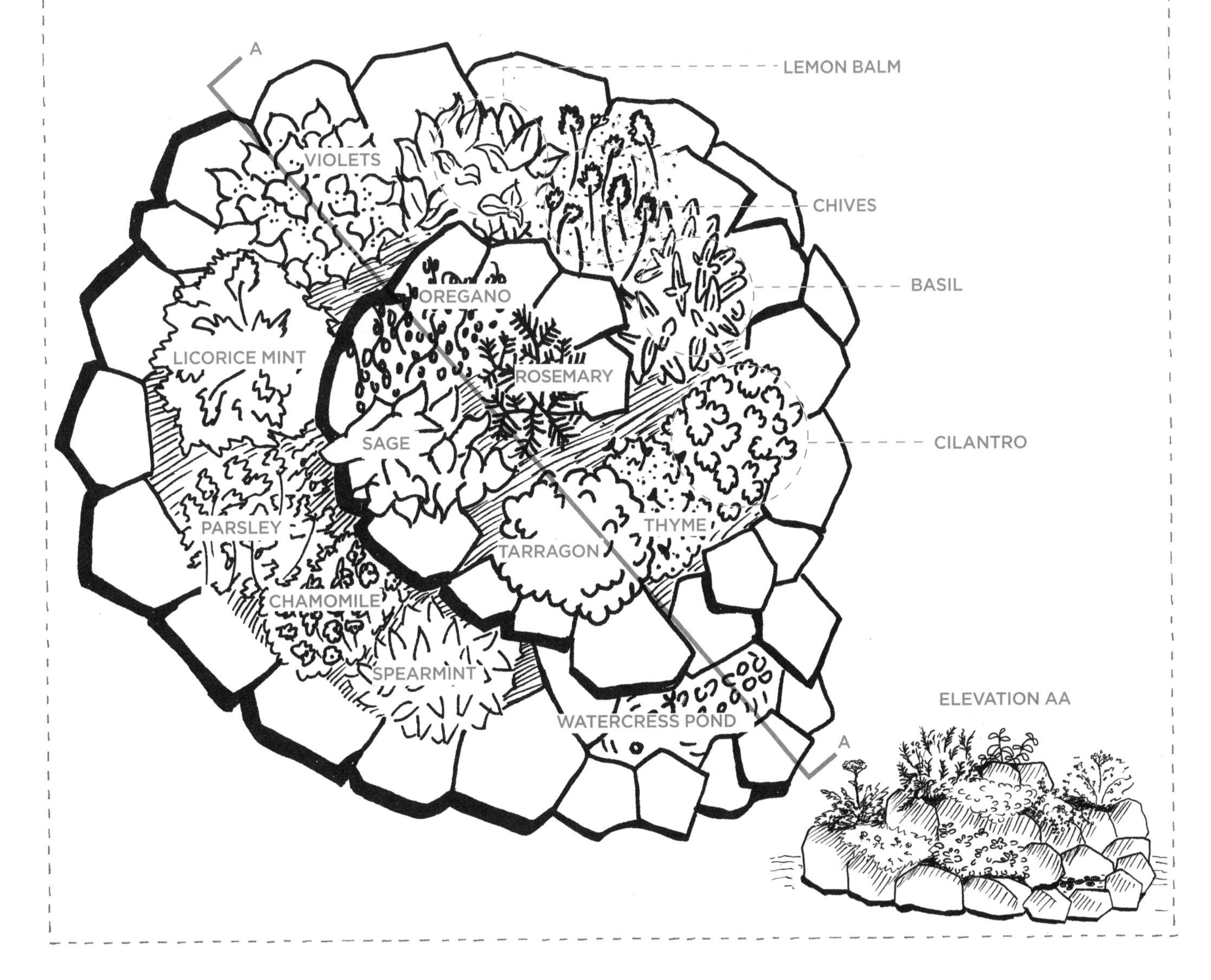

Basil

Although there are many varieties of basil, I like the traditional Italian 'Genovese', a deep-green, large-leafed variety. It can be harvested almost continually, and does well in both herb spirals and containers. To prolong the harvest, keep pinching leaves back a few times a week.

Chervil

This anise-flavored herb is often used in French cuisine. It is easy to integrate into the permaculture garden, because it can grow in full sun or partial shade. It needs the same amount of water as salad greens, so you can interplant it with lettuce and other salad mixes.

Chives

Chives are a perennial herb from the onion family. These grasslike plants can be tucked into the border of vegetable beds or grown in containers. The edible spicy purple flowers attract butterflies and bees. They also make pretty garnishes that add an onionlike flavor to salads.

Cilantro and coriander

The names cilantro and coriander are used interchangeably, but they refer to the same plant, *Coriandrum sativum*, at different stages of growth. This is an herb that people tend to love or hate (apparently some people have a genetic trait that makes cilantro taste like soap). If you want leafy cilantro, you will need to seed successively throughout the growing season; otherwise, it bolts and becomes coriander. This plant has a fairly low germination rate and short seed life, so it's best to save the seeds and replant, however 'Slo Bolt' is a variety true to its name.

Christopher's Garden: CREATIVE CONTAINERS

If you have limited space, or want to maximize your efficiency, reusing containers is a great way to gain a little extra garden space. Instead of buying something new and expensive, try to find something that would otherwise end up in the waste stream. Herbs make ideal plants for containers, and you can keep them on the deck or porch so they are close to the kitchen.

I use old nursery pots, salvaged clay pots, olive oil tins, oak barrels, and food grade 55-gallon drums (never reuse a container that has contained chemicals). If you cannot find enough from your own recycling bins or those of your neighbors, check your local recycling center, salvage yard, or thrift store. The Thrift Shopper website (thethriftshopper.com) is a nationwide directory of more than ten thousand charity-run thrift stores.

I try to keep some cilantro going in shadier spots during the summer, but because they do bolt so quickly, I often pull out the entire plant and add the leaves and stems to a stir-fry or soup pot. I always make sure to keep some of the choicest specimens for seed, though.

Dill

Dill is another classic summer herb, and in Mediterranean climates, it can be planted year round. It's a tall, delicate plant graced with yellow flowers at the end of umbrella-like flower heads. Dill is related to celery and parsley and, like all members of the carrot family, it attracts beneficial insects. Butterfly larvae even like to use this as a host plant, so I always leave some dill in the garden for the swallowtail butterflies to eat.

Lovage

Lovage is a traditional herb that may be unfamiliar to many gardeners but deserves to be grown for its multipurpose nature. The leaves have a flavor much like celery with a hint of anise, and can be used in soups, stews, and casseroles. You can also use the dried seeds much like celery seed. It's a perennial herb that will often reseed, and can be dug up and divided to make new plants. Lovage can grow quite large, so it's best used in the herb layer, but it can reach 6 feet if not harvested regularly.

Oregano

Another Mediterranean cooking herb, Italian oregano is a good addition to your herb garden. Choose the right spot when planting oregano, since it can take over half the herb spiral if you fail to harvest it regularly. Cut back your oregano hard a few times during the growing season and toss the clipped stalks on an herb rack to dry. Once dry, prepare oregano for storing by putting the stalks in a paper bag. Rub the outside of the bag to separate all the leaves off the stems and branches, then gently rub the dried leaves through a seed screen and package in glass jars or plastic bags for storage.

Chives in flower.

Parsley

I'm a big fan of 'Italian Flat Leaf' parsley because it's an easy-to-grow variety, with consistently large yields that seem to regrow quickly after harvesting. Parsley is a useful seed-saving herb that can easily become a reseeding annual or biennial in your garden. Parsley also contains vitamins and iron and is remarkably hardy, often continuing to grow even after snowfall.

Rosemary

Fragrant rosemary grows into enormous bushes in warm climates, making dramatic hedges of dark green, fragrant leaves with vibrant blue flowers in winter. It is a magnet for bees and other pollinators and is easy to care for and drought-tolerant. The upright varieties are too large for herb spirals, so plant them on their own, but within easy reach of your kitchen. Creeping rosemary can be used in an herb spiral, as ground cover, or draped over a rock wall. In cold-winter climates, take rosemary cuttings inside during the fall, root them and keep as a window or greenhouse herb.

Sage

This drought-tolerant, aromatic herb works well in the top south side of the herb spiral, where it gets full sun and good drainage, but it can also be used in the low shrub layer in other areas of your garden. Honeybees love garden sage and it's a treat to have this earthy herb available for fall dishes like stuffing and pumpkin soup.

Sorrel

Sorrel is not often found at the grocery store because it doesn't ship well. But it's a low-growing perennial herb that can tolerate part shade, so it's easy to integrate into your herb spiral or vegetable beds. In my mild

Flat leaf parsley.

climate, it grows year round, but in cold-winter climates, it will die back in winter and re-emerge in spring. Use sorrel in soups and sauces; it has a slightly acidic tang that can be tempered by cooking with cream.

Tarragon

Many people are familiar with the licorice-like flavor of tarragon from its use in French cuisine (in favorites such as Bearnaise sauce) and as an additive to flavored vinegars. It's a perennial in coastal climates and should be planted in the lower south of the herb spiral, because it likes more moisture than many other herbs.

Thyme

Thyme is a great landscaping plant for using edges and valuing margins, because it is so easy to tuck into small corners as both a low-growing ground cover, or as edible greenery cascading down rocks walls. When thyme is producing copious amounts of tiny white or purple flowers, I have seen so many honeybees that I think the plants will run out of pollen—but the bees keep coming. Like rosemary, thyme is easy to take cuttings from and bring inside during winter. There are numerous varieties with flavors of caraway, coconut, lemon, lime, and more.

HERBS FOR TEAS AND REMEDIES

Many herbs can be made into teas, and in a number of cultural traditions these teas are considered to have medicinal properties. Most people are familiar with chamomile tea as a sleep aid, but may not know that mint tea can help soothe a stomachache, and tulsi tea is said to help boost the immune system. It's also satisfying to have some of these herbs handy, whether fresh or dried, to make tea from your own garden when guests come over.

Teas can be made from fresh or dried herbs. If you use fresh herbs, tear or crush the foliage to release the essential oils. In some cases, the flowers make the best tea—chamomile is an example. In general, you'll need twice as much of a fresh herb to get the same intensity of flavor as dried herbs. Try mixing different herbs together or combine with edible flowers, berries, and spices.

Licorice mint (*Agastache foeniculum*)

Licorice mint is also known as Korean mint or anise hyssop, and it's one of the most showy herbs in the garden, with large green serrated leaves and lofty purple spires of flowers. The variety 'Tutti Frutti' has beautiful raspberry-red flowers that attract hummingbirds and beneficial insects. Licorice mint works well as part of the herb spiral because it's a compact herb that won't spread from underground rhizomes like other mints. Save seeds from this annual and plant again next spring.

Catnip (*Nepeta cataria*)

Easily cultivated in any garden soil, catnip makes a nice nighttime tea. It is said to repel mosquitoes, so plant it around your seating areas in zone 1. It's not an appropriate choice for an herb spiral unless it's a big spiral—as a member of the mint family, catnip can spread aggressively.

German chamomile (*Matricaria chamomilla*)

A cup of chamomile tea has a soothing quality that can help me get to sleep, so I always try to grow some German chamomile in my garden. This shrub-sized plant produces lots

Thyme in flower.

of flowers, which you can gather by the bagful for drying. While I have seen a few lawn alternatives to German chamomile, such as Roman chamomile, they take more work to keep watered and weeded. Chamomile will reseed itself, but save some seeds just in case they don't come up. An added bonus is that it's a really pretty plant that smells great—and the bees love it too.

Lavender (*Lavandula* species)
Most people recognize the beautiful flowers and fragrance of lavender as a plant that has soothing and calming properties, yet they might not know the flowers are edible. You can also use the leaves in cooking, especially for roasted meats and even in baking. This amazing insectary plant likes good drainage, so if you are using in an herb spiral, put lavender on the top toward the south side. There are many beautiful varieties of lavender, so it makes a great plant for mixed borders and works well interspersed with rosemary. Dried lavender flowers can be put into sachets for drawers and linen cupboards.

BELOW
Lavender and poppies mingle together in this hot summer bed. They attract insects that help pollinate the grapevines behind.

OPPOSITE
Harvest fast-growing members of the mint family often to keep them in check.

Lemon balm (*Melissa officinalis*)
Lemon balm is in the mint family and is shade tolerant, but it doesn't spread from rhizomes like most mints. Instead, it will reseed, and new clumps will volunteer in different locations each year. Learn to recognize the seedlings of this versatile herb and use the leaves to make tea.

Mint (*Mentha* species)
I keep a small container of spearmint on the deck for easy access so that I can snip a few sprigs to add to lemonade, meat dishes, and pesto. My favorite mint is peppermint and we have a large patch of it at Merritt College as a ground cover around some plum trees. In addition to making an excellent tea herb, mint is great bee forage and gives honey a superb flavor. It's nice to have mint around if you have enough space for a scented garden. Mint can take part-shade conditions, but be careful where you plant it because it spreads aggressively and can take over a bed quickly.

Mugwort (*Artemesia vulgaris*)
There are Chinese, Japanese, and North American species of mugwort (also known as woodworm). It is used in many herbal traditions as a sleep aid and is said to have various healing properties. Traditionally, rubbing mugwort on the skin is said to soothe or prevent the rash caused by poison oak. It's a tall plant, not ideal for herb spirals or containers, but better used for transition areas from low fruit trees to smaller herbs. Tea can be made from fresh or dried leaves.

Red clover (*Trifolium pratense*)
Red clover is one of the permaculture star plants. It is a nitrogen fixer that can be grown

in quantities as a cover crop. It can be tucked in between other plants, and has beautiful deep pink flowers that attract pollinators and beneficial insects. You can sow it as a ground cover or in a big herb spiral where it has room to spread. The tea is made from the dried flowers, and considered to be a blood cleanser and overall tonic.

St. John's wort (*Hypericum perforatum*)
Known for its anti-depressant qualities, this yellow-flowering herb can be made into a tea that is thought to be more effective if drunk in the sun. It's really a spreading ground cover that can be grown only in a large herb spiral. Some varieties are small, shrublike plants that work well interspersed with other perennial herbs like lavender.

Tulsi (*Ocimum tenuiflorum*)
My wife is from India, where tulsi, also known as holy basil, is a sacred herb used in Ayurvedic medicine. It's a perennial that sometimes grows wild in India, but in our mild climate it behaves like an annual. It grows well during warm weather but expires in cold, wet conditions. Like other basil species, unless you are growing it for seed, continually cut back the flowering buds to keep the plant in leaf production. Honeybees and hummingbirds love to drink the divine nectar of this herb.

EDIBLE FLOWERS

Edible flowers deserve a place in your herb spirals as well as in garden beds. They encourage beneficial insects and bring beauty to the garden. If you let flowers go to seed, they surprise you by coming up in unexpected places. You can either leave them there, or dig them up and give them a new home.

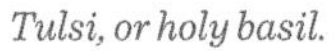

Tulsi, or holy basil.

Honeybees

Honeybees are essential for pollinating most flowering plants, including fruit and nut trees, vines, and many annual and perennial crops. They also make honey, which is a delicious crop in itself. In permaculture, there are different approaches to honeybees in the garden. First is a hands-off approach. Simply let a colony of bees establish a hive in the hollow of an old tree in zone 5, and never try to get their honey. This requires no work and you still reap the benefits of pollination.

The second option is for those of us who love to eat honey, can lift heavy objects, and are willing to get stung a few times along the way—and that is to keep beehives somewhere in or near your garden.

There are also two basic types of hives, both of which are used at Merritt College: the traditional Langsdorf stackable boxes and the Kenyan Top Bar beehive. The Langsdorf gives more consistent harvests and is more production-oriented, while the Top Bar resembles more how honeybees make their nests in the wild.

Although some beekeepers prefer to suit up from head to toe and smoke out their honeybees to harvest the honey, I see beekeeping as an opportunity to practice a slow food approach. I prefer not to smoke out the bees, but go in gently and slowly as though I am doing tai chi, wearing only an inexpensive veil and a bee suit with gloves.

There are many benefits to eating honey from your own hives. Some people believe that eating a spoonful of local honey every day can build immunity to regional allergens. What's certain is that honey tastes delicious, and artisanal honey can be sold for as much as $10 to $15 per pound. With a single hive, I can get nearly 100 pounds of honey per year, so that represents a significant savings. We have started to substitute honey for sugar in baking, and have tried to make fermented honey drinks such as mead, wine, and kombucha. Honey is also a perfect crop for sharing with neighbors and friends.

The Langsdorf style bee boxes at Merritt College.

Calendula.

Edible flowers are part of the herb layer in the edible forest garden. Some edible flowers provide many yields. For instance, both the tubers and flowers of daylilies are edible, they are dynamic nutrient accumulators, and they provide natural weed control. But remember, not all flowers are edible! For example, foxglove contains a powerful compound that is used in heart medication but is poisonous if ingested. Do your homework to find out which flowers in your garden are edible and always err on the side of caution.

Borage (*Borago officinalis*)
Borage, with a flavor reminiscent of cucumber, is a good edible flower choice because it makes so many flowers and reseeds itself, producing flowers for a long time. It's a dynamic accumulator that can outgrow its space, so I occasionally pull out a specimen by the roots and add to the compost pile or spread it as mulch.

Calendula (*Calendula officinalis*)
My younger daughter already recognizes calendula as a flower that is safe to snack on in the garden. In addition to its culinary use, calendula can be infused in olive oil to make a therapeutic skin ointment. The calendula plant dies back and reseeds several times a year, so it makes a good plant for the edges.

Chrysanthemum (*Chrysanthemum coronarium*)
The edible chrysanthemum, also known as chop suey plant or shungiku, has strikingly beautiful edible flowers and leaves that are used in Chinese cooking. It's a large plant—reaching 3 to 4 feet tall—so not a good choice for an herb spiral; it's better to intersperse these flowers in your vegetable beds.

Daylilies (*Hemerocallis fulva*)
Daylilies are perennials that are truly carefree plants, coming up year after year and requiring no maintenance. Both the buds and the flowers of daylilies are edible. Because the flowers last only for a single day, pick and eat them right away, adding to salads or dipping in batter and then frying them. The buds can be cooked like green beans, steamed, or added to stir-fries. Dried

daylily petals are known in China as golden needles and can be added to soup.

Nasturtium (*Tropaeolum majus*)
A cheery ground cover with neon-colored blooms, nasturtiums can be grown in a container, on a trellis, or mixed in with other vegetables. In Mediterranean climates, it grows all year round, adding color to fences and gaps in the border. The edible flowers are good for salads and deserts, the spicy young leaves are eaten fresh, and the prolific seedpods can be pickled like capers. It truly does obtain a yield in the permaculture garden.

Pansy (*Viola* species)
This beautiful ornamental comes in a huge range of colored and multicolored varieties, all with sweet-tasting edible flowers. It makes a cheery border plant for vegetable beds, and the flowers look lovely on iced cakes and cupcakes. Pansies also make good container plants.

Rose (*Rosa* species)
The queen of the flower garden comes in endless sizes, ranging from miniature shrubs to scrambling vines. The flower shapes are just as varied, although all are edible, which make sense as the rose family also includes apples, blackberries, pears, and raspberries. The petals can be used to make rosewater or added to lotions and bath products. Rosehips can be made into jelly, jam, wine, or tea.

Scented geraniums (*Pelargonium* species)
This is a native South African drought-tolerant plant that can be grown in containers, although some types are quite large—mint geranium can be grown as a ground cover and rose geraniums are large shrubs. Scented geraniums generally like full sun

Nasturtium.

but a few types can thrive in partial shade. Many have leaves with exotic fragrances like apple, apricot, chocolate, mint, and rose. Citronella geraniums can help to repel mosquitoes. These plants are tender perennials, so bring them indoors during cold weather. You can easily multiply your crop by taking cuttings in fall.

Violets (*Viola odorata*)
Violets prefer the moist shady placement on the lower north side of herb spirals. They don't need maintenance once established, so these cute and colorful flowers make a great addition to the permaculture garden. The petals may be used fresh as garnishes or mixed in fruit salad, crystallized for decorating, or dried for addition to savory dishes.

Grains

Most edible gardeners don't consider themselves to be grain farmers, and most of us don't have the acreage for growing meaningful quantities of grain. But it's still a satisfying endeavor to have a small grain plot and be able to make a few complete meals out of your own crops. In fact, there are both traditional and nontraditional grain crops that can be grown in the home garden. 'Burgundy' amaranth and 'Faro' quinoa make a striking splash of color as a garden backdrop, and with bigger spaces, you can plant winter wheat, which brings beautiful waving movement to the landscape. All grains make protein-rich seed heads and the non-edible stalks have the added benefit of providing

Christopher's Garden: FLORAL PERMACULTURE

When I was coming up with a vision of our home garden, my wife, Runa, an integrative physician, wanted to make sure I included flowers and herbs for their healing properties.

Our flower crops the first year were cosmos, sunflowers, and sweet peas, but after the second season, we had flowers integrated throughout the garden. The herb bed was full of flowers, including purple coneflower, white sage, and St. John's wort, with strawberries planted on some of the outside edges. A couple of years later, we had the pleasant surprise of a few varieties of poppies that showed up on their own. I weeded out the ones that were taking up good spots for vegetables, but left many red, golden, and pink poppies along the margins. Permaculture encourages us to plan our garden space, but also leave room for nature's contributions, and when they come in the form of beautiful flowers, they are most welcome!

These days, we've long given up on having a single bed for flowers, and my wife is happy to see an inviting and productive garden with something always in bloom. The flowers draw insects and birds who love to have nectar to sip and pollen to eat. Now, in addition to enjoying more traditional flowers, we also appreciate the blooms from our edible crops, such as the 'Costoluto' tomato, herbs like chives with their lavender flowers, and flowering trees like the heirloom apple 'Hidden Rose', which makes the largest and pinkest flowers I've seen on any apple tree.

really good mulch. After threshing, the stalks of amaranth, barley, or wheat can be laid down to protect the soil. Then the next crop is fed by the rotting biomass of the previous crop.

If you have limited space and really want to grow grains, ask your neighbors if you can use some of their land, or try to find a community garden plot. If you can find a large enough space and some other like-minded growers, then band together and form a co-op to share the labor, tools, seed, and fun of growing grains. It helps to have a lot of people for the harvest, drying, threshing, and winnowing of grains. If you have large drying space, like an underutilized garage, rafter space, or attic, then grains can be dried for later processing or, alternatively, fed to small livestock like chickens.

Alley cropping is a permaculture technique for growing a mix of grains and legumes between orchards and blocks of food forests. The alley is a narrow strip between larger blocks of planting, similar to an alley between houses. Just leave the equivalent of a few extra rows of space between your fruit trees to put in the grains. With grains you need sufficient quantities of yield to make it worthwhile, so alley cropping is a way to increase the yield.

A ground cover of clover can be used along with grains to fix nitrogen and enrich the soil. Plant the grains and the clover at the same time, or if the clover is already established, hoe out rows within the cover crop for the grains.

Amaranth

This grain was once a staple of the Aztec people, and is now gaining popularity as a small-scale grain. Amaranth is also an obvious permaculture choice, as it self-sows, is a strikingly beautiful plant, and is nutritious. The grain is high in lysine, an important amino acid not found in other grains, and the young greens contain more calcium than milk.

Amaranth is a tall annual grower that likes warm weather. If you are growing amaranth for seed saving in wetter and colder climates, the whole plant can be harvested in late summer and hung upside down in a garage with a tarp underneath, for threshing and winnowing later.

Barley

Barley is a cool-season grain crop that is especially useful as part of a cover crop mix. Its large root system works its way into the soil, providing tillage. You need a large bed (at least 4 by 12 feet) to yield enough for a food crop.

Buckwheat

This grain is traditionally used as a summer cover crop. If you have enough space, you can grow some for about six weeks to improve the soil and attract bees. Then, cut down and mulch the buckwheat, and replant the beds with late-season broccoli or kale.

Flax

Flax is not only used for fabric (linen) and for its ornamental red and blue flowers, but

Chickens

A great reason to grow grains is to feed your chickens, which are exactly the kind of low-maintenance, full-circle element that permaculture calls for. I have chickens and ducks in my backyard, and they add a lot more to the food forest than just fresh eggs. Chickens can be run in fallow beds, acting as chicken tractors to eat insects and weeds, and provide natural fertilizer, which they aerate themselves as they scratch the dirt. In mulched areas, there will be a lot of worms, sow bugs, and beetles that the chickens snack on all day long. My daughters and I make periodic forays into the neighbor's unkempt yard and pick jars full of brown garden snails to feed to our chickens.

Whenever you add to your compost pile, it's a good time to clean out the roost where the chickens sleep at night. The bedding and manure are full of nitrogen and phosphorus that helps heat up the compost. At least once a year, I dig out the chicken pen to the native soil level, usually over 18 inches. I add that to the compost pile and then, after it has decomposed, put it back into to the garden beds in spring.

Eggs are the main reason we make the effort to keep chickens, and they taste so much better when the chickens have a wider diet of more than just grains. The egg yolks are a richer yellow-orange color, have a higher nutritional content, and I believe they are safer than eggs from factory-farmed chickens.

Chickens need a secure, warm place to sleep at night and enough room to run around the food forest during the day. You can let them into all areas of the garden except where you have young seedlings. It's especially beneficial to have them perform slug and snail patrol in the evening. Chickens are an essential part of our permaculture garden, and my older daughter and her friends spend a lot of time sitting on a bench outside their coop, feeding them grains, greens, and insects from the garden.

Chickens provide an output of fresh eggs, thanks to the input of snails provided by Gitanjali.

'Burgundy' amaranth is one of my favorites because of the brilliant deep red color of the seed stalks and leaves.

the oil is highly prized. When it first emerges, it's a delicate-looking plant, but eventually it turns into a 2-foot tall, fibrous, and wispy flower that can be interplanted among vegetables and fruit trees.

Millet
Millets are annual grasses that prefer warm temperatures, and there are many drought-tolerant types, including foxtail millet and proso millet. Millet makes a fluffy rice-like dish, and it is also a primary ingredient in birdseed and livestock feed.

Popcorn or flour corn
Unlike sweet corn, these varieties of corn are only consumed after being dried. The varietal names often reflect the crop's history as a Native American staple: 'Cherokee Long Ear' (popcorn) and 'Blue Hopi' (flour). 'Two Inch Strawberry Popcorn' makes a whole bowl of popcorn from just one deep-red, strawberry-shaped ear. It looks like a mix between a porcupine and a corn plant, so I use it for decoration in the fall as well as for popcorn.

Quinoa
Quinoa is a South American pseudo-grain from the Andes, where it grows above 8000 feet. In the Bay Area, quinoa can be planted year round, and I like 'Faro' quinoa as it is very productive in small spaces. In cold-winter climates, you can start quinoa in a greenhouse and then transplant into the garden in the late spring. Quinoa likes full sun and is drought tolerant.

Growing plants from seed you have saved from the previous years' plantings means you can select those edibles that do best in your garden.

6 GROWING FROM SEED

Seed or Starts?

IF YOU HAVE NEVER tried it before, growing vegetables from seed may seem intimidating, but it is actually relatively easy—at least for the majority of vegetable crops. All you need are seeds, good potting soil, cell packs (seedling containers with individual cells linked together), sunlight, and water. It's definitely worth growing plants from seed, because you can save hundreds of dollars, and you can try a much wider range of varieties, including rare and heirloom varieties that are not commercially available as transplants. Growing from seed means that you produce no waste and offers the opportunity to see your garden in terms of renewable resources—your crops themselves become the source of the next year's produce.

Many annual and some perennial edibles can be direct-sown in the ground, but there are advantages to growing your young plants in cell packs for later transplanting into the garden. The seedlings are bigger and stronger in cell packs than when sown directly in the field, so they survive insects and birds better. You do not have to do as much thinning of overplanted seedlings as when direct seeding. In cold climates, you can start many warm-season crops in the greenhouse or under glass for transplanting when the soil has warmed and all danger of frost has passed.

Direct Seeding

Direct seeding is best for crops like carrots that have a long taproot, or those that mature quickly, like radish, lettuce, annual herbs, and salad greens.

Field planting a crop of peas saves time, money and nursery space. If you have a large planting, an Earthway seeder is a wheeled

Planting out young vegetable starts at Merritt College.

These young starts in six-packs will be ready for planting out in the garden in several weeks.

tool that can evenly seed a bed in drills, spacing the seeds anywhere from 1 to 12 inches apart. This avoids the need to thin crops that have been over-sowed. You don't need precision planting for all crops, though. Grains, dry beans, and cover crops like barley, buckwheat, clover, fava beans, rye, soybeans, and vetch can be directly broadcast by hand.

Growing from Starts

Seedlings are best raised either in a greenhouse or on some kind of nursery table where you can keep an eye on them and protect them from slugs and snails. Depending on the crop, your nursery table may be in full sun, part sun, or shade. Most vegetable starts will need at least 6 hours of sun a day; exceptions are ferns and other woodland natives, as well as edible shade plants like French sorrel, shungiku (edible chrysanthemum), and nettles. I typically grow starts in flats comprised of eight six-packs, totaling forty-eight individual plugs. Plant seeds in small batches unless you are trying to grow food on a large scale to sell or trade, or you plan to give away starts.

Start seeds in a good-quality sterile potting mix. Buy the best quality organic (and preferably local) potting mix you can find

Large-seeded crops such as beans, corn, peas (shown), and squash are also typically direct-sown, especially if you are doing large field plantings.

and then add 10 percent by volume of worm castings. Always label your cell packs with the type of seed, the variety, date, source of the seeds, and any special considerations. When planting a flat, begin at the top upper-left corner and work your way down filling the empty cells with seeds (just like reading a book). Generally, I just drop two or three seeds into each cell. For large seeds like squash you only need one or two seeds, and for hard-to-germinate seeds like celery or cilantro, you may need three or four. Remember that if too many seedlings come up, you will have to thin them out which is more work.

Press the seeds into the potting soil and cover with an additional sprinkling of potting soil, making sure to just cover the seeds and not bury them too deeply, or they will rot. The general rule is to plant the seed as deep as its own diameter. Large seeds like beans can be planted one inch deep and tiny seeds like basil are planted less than an eighth of an inch deep.

Keep your seedlings well-watered, and transfer them to larger pots if they outgrow their cell packs. When you are ready to transfer them to the garden, choose an overcast day if possible. Water in the young plants well, and protect them—but don't smother them—with mulch.

SEEDBALLS

Grains, wildflowers, cover crops like clover, and some vegetables can be pelletized in a mixture of clay and compost known as seedballs. Seedballs are an ancient technique; known as earth dumplings, they were reintroduced by Masanobu Fukuoka. They are easy to make and protect the seeds from birds and rodents. Simply mix three parts clay, one part compost, one part seeds, and a little water, and mix until you can roll balls that stay together. Dry them outside and then toss them wherever you want in the garden. The rains will melt down the clay mix, and the seeds will come up on their own. Children love to make and scatter seedballs, and you can support a little guerrilla gardening by making sunflower seedballs and encouraging kids to throw them into vacant lots or their own gardens.

SEEDBALLS CONTAINING CLAY, COMPOST, SEEDS, WATER

MELTING SEEDBALLS

NURSERY TABLE

I like to grow my starts on a nursery table so I don't have to bend over too much! In very small spaces, a nursery can be set up on a salvaged wooden table, old metal roofing set on barrels, or planks laid atop cinder blocks (shown)—really, any secure surface will do. The nursery does not need to be within the garden, but can be placed on concrete or other hardscaped areas. The space under the nursery table can be used for shade-loving plants—I grow shiitake and oyster mushrooms under my tables in spent mushroom blocks.

Seed Saving and Storing

People have been saving seeds for thousands of years. However, in the last hundred years, much of the seed-saving knowledge in our culture has been lost, and this has also led to a significant loss of biodiversity. In pre-agricultural communities, indigenous people used on average 1200 species of plants. In 1900, there were 1500 diverse plants feeding most of the people in the world. Today, 90 percent of the world's food comes from thirty different plants, and of these, three quarters of the world's food comes from only four crops: corn, rice, soy, and wheat.

In the early twentieth century, there was 98 percent more crop diversity than we have today, which made us more ecologically resilient. In the U.S. in 1970, hybridized corn carrying an identical gene succumbed to a single disease affecting 50 percent of the corn crop in the South and 15 percent nationally, at the cost of a billion dollars. Plant scientists went to Mexico, the origin of corn cultivation, and found landraces—native varieties of corn. They selected the landraces with the best disease resistance and bred those qualities into the next generation of hybrid corn. The lesson from this is to follow the knowledge of those indigenous farmers who would never have planted just a single corn variety. This is why permaculture encourages saving and trading seeds and the planting of polycultures.

WHY SAVE SEEDS?

Seed saving is one of the best ways to practice permaculture. Through seed saving, you learn to select the best crops from each year's yield, while saving money on next year's garden. Seed sharing provides

I try not to overseed when sowing into cell packs, because that just means more seedlings to be thinned a few weeks later.

opportunities to expand your plant knowledge, grow some new varieties, and meet other people with similar interests. Though it may seem intimidating at first, I urge you to try seed saving.

The act of saving seeds demonstrates several permaculture principles. A good example of obtaining a yield is the fact that a single ounce of dried kale or collard seed can contain seven thousand seeds. That's a lot of seeds! As you can't possibly plant and grow them all, it ensures a level of generosity and encourages community networks as you trade with other seed savers. Share your seeds with friends and just imagine all the greens you will have helped grow.

Masanobu Fukuoka also teaches us a lot about seed saving, polyculture, and observing natural rhythms. Visitors to his farm had a difficult time finding where he grew his vegetables, because the fields between the orchards often looked overgrown and fallow. But Fukuoka's trained eye saw daikon radishes going to seed amidst all the other wild plants. When you begin your journey toward natural farming—the root of permaculture—you will see the importance of letting things go to seed, and notice the connections from season to season.

When you grow and save your own seeds, you develop crops that are well suited to your climate and growing conditions, and more resistant to pests and disease. In addition, seed saving allows you to select for desired characteristics including flavor, taste, color, smell, texture, shape, juiciness, and tenderness. As an added bonus, saving heirloom seeds preserves genetic diversity, which promotes earth care.

You want to make sure that you always save seeds from the healthiest plants with the desired characteristics of that particular variety. In essence you are staying, "Next

BOTTOM LEFT
Saving seeds from fava beans and arugula is a good opportunity to share a task.

Get to know the wide variety of seeds from the edible crops in your garden, such as kale (TOP LEFT) *and sunflower* (FAR RIGHT).

year I want more plants like you." Look for plants that are resistant to diseases and insects; so, for instance, if some lettuce plants are being eaten less by slugs, allow them to go to seed so you can grow them again.

REGIONAL ADAPTATION

Regional conditions affect how you save seeds and which types to save. In my garden it can be challenging to grow a large tomato like 'Brandywine' because we don't get enough summer heat, so it's better to grow cherry tomatoes, like the heirloom 'Chadwick Cherry'. But if you live in a different climate, you will be selecting for other traits like insect resistance, drought tolerance, short seasons, high elevation, or soil type. In climates with high humidity and heavy rainfall, for instance, diseases like early and late blight are common in tomatoes, and the seeds most resistant to these conditions should be selected.

This past winter, I observed some 'Chinese Giant' red mustard during a cold snap—two of the plants were damaged by frost, but two survived without any damage. Of course I saved seed from the survivors. By saving seeds from the hardiest, tastiest plants, you're letting nature help you with her gift of natural selection.

HYBRIDS AND HEIRLOOMS

Some basic botany can teach you how closely plants are related and will help your seed-saving efforts. Take the time to learn the family, genus, species, and variety of your plants. Botanists have classified plants with the same flower patterns and structure in botanical families. Within each family are one or more similar genera (the plural of genus). So, for instance, the family Asteraceae (the aster, daisy, or sunflower family) contains many plants that have composite daisylike flowers, such as lettuce (genus *Lactuca*), marigolds (genus *Tagetes*), sunflowers (genus *Helianthus*), and many more. Different genera will not usually breed with each other—marigolds will not breed with sunflowers. Genera are further divided into species, indicated by the second Latin name, such as *Lactuca sativa*, *Tagetes patula*, and *Helianthus annuus*. Unlike different genera, two plants of the same species do breed with each other. When two genetically similar plants produce offspring that are identical to the parents, the plants are said to be true to type, and the seeds are referred to as open-pollinated.

In contrast, hybrids are first-generation offspring that result from two dissimilar parent plants. Most are produced by large seed-breeding companies. These plants often have hybrid vigor—compared to the parent plants, they may be bigger, have better disease- and pest-resistance, and have more uniformity. Though these sound like ideal traits, unfortunately, it can take thousands of plants to make a hybrid and the rest are rouged (killed off), because they don't have the right characteristics. Breeding hybrids is an expensive and time-consuming process that most home gardeners cannot attempt.

Hybrids also require the gardener to purchase new seeds each year, because hybrid seeds do not breed true to type. I recommend you stay away from hybrids for seed saving (labeled hybrid or F1). I once saved seeds from a 'Sungold' cherry tomato, which wins many taste tests. I ended up with a 12-foot-long vine that made bitter and thick-skinned micro fruit. So feel free to use hybrids in your garden, but don't plan on saving seeds from them.

Open-pollinated or heirloom seed varieties, on the other hand, have been grown for at least fifty years and handed down from one generation of gardeners to the next. It is from our grandparents' seeds that we continue this tradition. So each season, open-pollinated seeds result in plants that will be very similar to the parents, but will also start to adapt to your local conditions.

POLLINATION

A seed is living potential and requires the right conditions to germinate, grow, and thrive. Once a seed becomes a plant, if left alone, it will eventually produce seeds itself in order to reproduce. For human life to take place, an egg needs to be fertilized by sperm. Pollination is this fertilization process in plants, and understanding how it works will help you grow plants true to type.

Some plants, like peas and beans, are self-pollinating, which means that male and female parts merge within each flower and don't require insects, wind, or human intervention. Seeds from self-pollinating plants are generally easy to save and a great place for beginning seed-savers to start.

Two other methods—cross pollination and wind pollination—are a little trickier. Cross-pollinators like squash need help for fertilization to occur, which can lead to unpredictable results for the seed saver. They have separate male and female flowers and depend on insects like honeybees and flies to get their pollen from the male flower to the female flower. Wind pollination, which happens with crops like corn, makes it difficult to reliably save true-to-type seeds, because corn pollen can travel for miles by wind.

For most plants, you should save seeds from at least six individual specimens, but if you have the space to plant more, then try to do so. Corn is the exception to most small-scale seed-saving programs. You need at least two hundred plants to save or else the plants will inbreed and the next generation will have smaller ears. Corn is likely to get mixed up with other varieties, so hand bagging (where you staple a paper bag over the male and female parts and bring in the tassels for pollinating) is needed for small plots.

Members of the same plant species often cross-pollinate and their seeds are saved in a similar way, which is why it helps to know your plants' botanical names. For example, if you plant broccoli, cabbage, cauliflower, and kale together, you may get beautiful vegetables this year, but may not be able to save the seeds of any of them for future crops because they are all *Brassica oleracea* and will cross-pollinate. The solution is to use distance, either physical barriers or staggered timing to prevent the plants from pollinating each other. Also ask your seed-saving neighbors which varieties they are growing so that you can avoid the risk of cross-pollination from nearby gardens.

If you have a big space or multiple gardens, you can avoid cross-pollination by putting enough distance between crops. Or you can have large plantings of beans, sunflowers, or other tall crops to separate different plants from which you want to save seeds. By having a physical barrier, you keep insects from visiting all of your crops at the same time and making crosses. Another example of a barrier can be a house, with different varieties planted on each side. Or you could grow some crops in your own garden and others in a community plot.

Let time work for you by rotating crops. The first year, for example, plan to save seeds only from kale, and the next year save seeds only from broccoli to keep a rotation going with all of the tricky crossers from this popular species. The seeds of most members of the *Brassica* family are viable for four or five years if stored properly. Another strategy is to stagger plantings. For instance, you could plant 'Red Russian' kale early in the season and another variety, such as 'Lacinato', two months later, so that their flowering times do not overlap.

Another option is an isolation cage. You can make your own cage by stapling row cover material to a wooden box frame made from lengths of 1- by 2-inch wood. Make the cage large enough to cover several plants at a time. You can also use row-cover material to bag individual flowers or flower clusters.

Processing Seeds

It's gratifying when you can roll up your sleeves and get down to the fun part: processing the seeds you've collected. In general, plant families can be divided into wet or dry processing categories (although nightshade family crops like tomatoes have more nuances). Dry processing typically involves letting the plant bolt (produce flowers) and then produce seeds, which dry on the plant. Wet processing is required for plants with seeds that won't dry on their own, like tomatoes, tomatillos, and zucchini. Once you get the hang of it, it's easy and you will know how to process the seeds you want to save.

DRY SEED PROCESSING

Dry seed processing is pretty basic. You collect the seedpods or seed heads when fully dried on the plants. When I first started gardening more seriously, I remember being very excited to see broccoli making bright green seedpods with tiny green seeds inside. My mother was visiting me at a community garden I helped start while in college, and I was eager to show her that now I knew where the seeds came from. She pointed out that I had to wait a little longer until the seeds were riper and black. If you pick seeds prematurely, most won't germinate as they are not fully developed.

By observing and interacting, you will learn how to tell when seeds are ready. The seedpods usually turn brown and are easy to crunch with your fingers. Watch for birds to start eating the kale seedpods, for example, and you will know it's time to bring in the harvest. Beans and peas should break cleanly and easily when properly dried; if

One of the reasons to grow plants like burdock is because they attract pollinating insects, butterflies, and birds, ensuring fertility in your crops.

you try to break a bean in half and it's rubbery, it needs more drying time. In shorter growing climates, bean and pea plants with unripe seeds can be hung up inside a shed or turned upside down with the roots still attached and dried.

THRESHING AND WINNOWING SEEDS

After further drying, you will need to thresh the seeds. Threshing is removing the seeds from the chaff—the rest of the seedpods or seed heads. Before starting, put down a tarp or blanket so you can easily clean up the seeds that miss the bowl. In general, break open large seedpods like beans with your fingers. For small seeds like kale, rub them between your hands over a large bowl in a back and forth motion, as though you are trying to warm your hands. For larger quantities of dried seeds, step on most seed heads and rub them on a tarp.

If you really enjoy seed saving, I'd recommend investing in a set of seed-cleaning screens. You can make these yourself or purchase them from a garden supplier. Not only are they handy for threshing, but you can use them for drying seeds (and herbs, too). Seed screens come with different mesh sizes to suit different size seeds, from very fine to coarse. Choose a screen that is just barely bigger than the seeds, and put an even finer screen underneath it. Shake the seed material through the bigger screen into the smaller one. Now find a screen that is slightly smaller than the seed, put the seed material into that one, and shake out the dust. What remains is the seed and a small amount of chaff which is lighter than the seed.

Winnowing is the separation of the seed from the chaff by blowing. Take the seeds and chaff that have come through the screen or have come free from the hand-rubbed seed heads, and dump the mixture from one wide and shallow mixing bowl to another while at the same time blowing gently on the seeds. You can also do this on a slightly windy day by pouring the seeds from one bowl to another and letting the chaff blow off. The viable seeds will remain in the bottom of the bowl, while the chaff blows off. If you throw the chaff into your garden, some missed seeds will always come up next season as volunteers. You can also put the chaff in the compost. After winnowing, take the clean seeds and put them in a glass jar or envelope. Label the seeds with the date, variety, and any special growing conditions. This will be important when you are ready to plant or pass them along to someone else.

At Linnaea Farm, British Columbia, where I apprenticed, they held a barn dance one evening to process a large quantity of barley. This is a good example of permaculture stacking functions: people having fun dancing on the barley while removing the seeds from the chaff. The next day we swept up the floor and winnowed the barley grain outside.

Seed screens make it easier to separate the seed from the chaff.

Sunflower or daisy family (Asteraceae)
Growing plants from the sunflower family is a sure way to get great seeds while also attracting beneficial insects. Members of this large family include artichoke, Japanese burdock, cardoon, chicory, endive, lettuce, sunflower, salsify, shungiku, and yacon. (An exception is sunchokes, which must be planted from tubers.)

Lettuce is a simple seed for the new seed saver to start with, because the plants are self-pollinated. Let plants come to full term and bolt, or produce a flowering stalk at the top, which takes many months. Some varieties have pretty yellow flowers and they all have white cottony seed heads. Lettuce seeds don't ripen all at once, so you can make several passes on them over a few weeks. Place a grocery bag underneath the plant, and rub the seeds with your palms into the bag to get the ripe seeds to fall off. Then winnow the seeds to get off the remaining chaff. Label your seeds with the variety, date, source, and any notes.

For large-flowered members of this family, such as sunflowers, seed collection is even simpler. The flowers are cross-pollinated by insects. Late in the season, cut off the flower heads when the seeds turn dark and start to rub off the flower easily. Hang the heads upside down in a warm dry place until the seeds can all be easily rubbed off. When the seeds are completely dry, bag and label them.

Legume family (Fabaceae)
This family includes peas and beans, including garbanzos, lentils, cowpeas, and peanuts. They are also some of the easiest seeds for beginners to save since they are self-pollinators (runner beans are an exception; they are pollinated by bees). I have a passion for collecting bean seeds, especially scarlet runner beans. It is a treat to take the nondescript bean pods and thresh them open to find little mottled pinkish red and black shiny seeds, much like a gardener's version of panning for gold.

Wait for the peas or beans to ripen and dry on the vine, usually about six weeks after the eating stage, and then thresh to open. Pea pods will be dry and brown, and if you shake them you will hear the peas rattling inside. Beans will be dry and yellow. Break open the pods and dump the seeds into one bowl and the pods into another to take to the compost pile. Children love to get into the treasure-seeking act and pull out the little gems from the old wrinkled pods. If you have a lot of beans or peas to thresh, put them in a burlap bag or pillowcase and step on it repeatedly until it has stopped crunching loudly. For a really large pea or bean threshing, set out two tarps and make a sandwich of the bean pods, then stomp away until you are ready to winnow them. Label your bounty with the variety, date, source, and any notes.

Peanuts are a special case because the legumes ripen underground. You have to dig out the entire plant when the leaves turn yellow. Hang the plants in a dry, warm place for several weeks. They need to cure slowly. After a few weeks, cut the peanuts off the plant and store them in the netted husk until ready for planting.

Onion family (*Allium* species)
The onion family, which includes common chives, garlic, garlic chives, leeks, onions, and shallots, produces highly ornamental

Lettuce that has gone to seed.

biennial flowers that range from white to purple. Saving seed is a bit more complicated, because they won't flower until the second year, after winter. In mild winter climates, bulbing varieties can be left in the ground to grow seed-to-seed, which means you can start with seeds and end up with seeds. In cold-winter areas, onions are grown root-to-seed, which means the best onions are dug up and carefully packed away in sand in a root cellar or non-freezing garage, and replanted in the spring. Then they are allowed to make the large flowering stage and produce dried seeds. Onion family flowers are unable to pollinate themselves, so you need a number of plants for pollination to occur with the help of insects.

Let the seeds dry on the plant and then harvest them. Thresh, ideally with seed screens, then winnow, bag, and label.

Goosefoot family (Chenopodiaceae)
This nutrient-dense family includes beets, chard, lamb's quarters, orach, quinoa, and spinach. In mild winter climates, beets can be left in the ground to grow seed-to-seed. In cold-winter areas, beets are grown root-to-seed, which means the best beets are dug up and carefully packed away in sand in a root cellar and replanted in the spring. Spinach is dioecious, meaning each plant is either male or female so you will never get seeds from a single plant. Aim to let at least six plants flower at once to increase the odds of pollination. Let the seeds dry on the plant, collect, thresh, winnow, and label.

Parsley or carrot family (Apiaceae)
This is one of the best groups of flowering plants to help create a balanced ecosystem. Beneficial insects can be seen in masses around the blooms, and you can often observe tiny parasitic wasps eating aphids.

This flowering leek is almost ready for seed collection.

The parsley family includes carrots, celery, caraway, chervil, cilantro (coriander), dill, fennel, parsley, and parsnip. If you have Queen Anne's lace in or near the garden, be aware that it is a wild carrot relative and may cross with your cultivated carrots. Carrots can show severe inbreeding depression within a generation, and become woody and inedible, so you need to save a lot of seed.

Many parsley family plants are biennial, so you must allow them a full year of growth and then collect seed the following year when they flower. To save seeds from carrots in mild-winter areas, you can leave the plants in the ground outside. In cold-winter climates, dig and store the roots (stecklings), then replant them the following season. Parsnips can usually overwinter in the ground even in cold-winter climates. Collect the seeds, which form on umbels, during the second summer. The umbels can shatter easily, so you may need to place a paper bag over the umbel to prevent the seed from spilling onto the ground. Let the seeds dry on the plant, collect, thresh, winnow, and label.

Brassica family (Brassicaceae)

This hardy family includes arugula, broccoli, Brussels sprouts, cabbage, cauliflower, collards, garden cress, kale, kohlrabi, mustard, radish, turnip, watercress, and many Asian greens. Brassicas do not self-pollinate and are all major attractors of beneficial insects. Most are biennial, and if planted in the late summer or early fall, will overwinter and produce seeds by the next spring. Remember to separate your brassicas if you want to save seeds, or allow only one type to flower.

If you are growing just one variety of each at a time, then arugula, radish, and watercress are easy ones to start with. Arugula won't cross with broccoli, and this is one crop with few varieties. Just wait for the flowers to dry into brown seedpods, then collect and shake some around on the garden where you want the arugula to come back.

For other brassicas, let the plants seed in the second season, and wait until the pods turn light brown. Don't harvest green pods, because the seed will be immature. If not all the pods are brown, you can pull up the entire plant and hang it in a warm, dry place until they mature. Brassica seed is tiny, so I recommend using seed screens before winnowing.

Grass family (Poaceae)

All grains are wind-pollinated, included those found in in the grass family such as corn, barley, oats, sorghum, and wheat. While corn is common, most gardeners are still not growing other grains and so you don't need to worry too much about cross-pollination. But if any of your neighbors are trying to grow an heirloom wheat at the same time as you, make sure they are planted far enough away so they don't cross-pollinate.

Corn readily crosses with different varieties up to a mile away, so it is unlikely that saved seeds will be like their parent plants. You need at least 200 corn plants to avoid

*Beetroot seeds are wind-pollinated and need to be kept separate from Swiss chard as they are the same species (*Beta vulgaris*) and will cross-pollinate.*

Christopher's Garden: SUCCESSFUL SEED SAVING

The plant I've been saving longest with the most success is a collard green. It is the variety 'Georgia Vates', and I was enthralled that the plant could get so huge—each leaf can potentially get bigger than a dinner plate. In our mild climate, it can almost grow like a perennial if cut back in the winter every year or two. I've grown it in most of my gardens and managed to save its seeds.

Ten years ago I planted some of these collards at Merritt College. We started on the top of the hill with the most marginal land, and now the collards have spread themselves around the whole hillside. Last year, we started a new area of the garden and dug in plenty of compost. This year, there is a healthy stand of collards in the swale above the trees. These were not transplanted: the seeds survived the composting process and sprouted on their own. They can also be scattered by birds.

This is the crop that yields the most and is the closest I've seen to the do-nothing farming technique. You just have to let the plants go to seed, harvest some for trading, and wait to see where they come up the following season. You can always move the seedlings if they're in the wrong spot. The main thing we do to ensure varietal purity is not to let the collards flower at the same time as other nearby brassicas whose seed we want to save. It helps that collards are more likely to breed true to type than others in this species.

genetic inbreeding, so it's not practical for most home gardeners to save corn for seed. If you have the acreage and want to do it, let the seeds dry on the plant, collect, thresh, winnow, bag, and label.

WET SEED SAVING

Wet processing is for those seeds that are imbedded in the wet flesh of the fruit and so don't dry out on the plant. In a natural environment, these plants would rot or ferment on the ground, and the seeds would sprout the following year. With wet processing, you allow this fermentation to take place in a container.

Nightshade family (Solanaceae)
The nightshade family includes cape gooseberries (ground cherries), eggplant, potatoes, sweet and chile peppers, tomatoes, tomatillos, and tree tomatoes. To save the seeds from these plants, you must allow the fruits to fully ripen before separating the seeds from the flesh. Potatoes are an exception, as they are normally grown from tubers.

To save tomato seeds, start with the ripest fruits of your favorite variety. You will need a plastic container for each variety—I use cleaned yogurt tubs. Squish the tomatoes into the container with your hands, and scoop out all pulp without seeds (you can use this to make salsa or sauce). Fill up the container, then put it outside, out of direct sun, and cover with a cloth to keep out fruit flies. Let it sit for about one week (depends on how warm it is) until a white fungus appears on the surface of the contents. This is a good thing, because you are trying to break down the gelatinous seed coat and the fungus will do the work for you.

Arugula has pretty white flowers that develop into seed pods. It reseeds readily and the seeds sprout early in spring.

Male and Female Flowers

How can you tell the difference between male and female flowers in the cucumber family? Typically the male flowers have a slender stem, and the female one has a swollen stem, much thicker than the male flower. Usually the male flowers appear first.

Male (LEFT) *and female* (RIGHT) *squash flowers.*

Next, empty the contents of the container into a sink for rinsing. The fertile seeds will sink to the bottom and the pulp and non-viable seed will rise up. Keep changing the water until it is clear and the cleaned seeds are at the bottom. Then put the tomato seeds on a seed screen or paper towel, and let them dry for a few days. Put the seeds into a paper envelope or a glass jar with a few holes in the lid so the seeds can continue to breathe. The last step is to label seeds with the variety name, date, location grown, and any special growing conditions.

Peppers don't need the same kind of treatment. Just cut open the ripe fruit and scrape out the seeds from the inside with a knife. If it's a hot pepper, wear gloves. Dry the seeds on a screen or paper towel, bag, and label.

Tomatillos are also easy. Simply toss the ripe fruit into a blender and turn it on low for a short time. Pour off the liquid pulp from the seeds (you can make salsa or put it into soups). Rinse the seeds and remaining pulp until the viable seeds are at the bottom. Then dry, bag, and label.

Cucumber family (Cucurbitaceae)
This family includes cucumbers, gourds, luffa, melons, pumpkin, and summer and winter squash. These plants require hand pollination. In the evening, tape up the male and female flowers, labeling the tape so you can identify them the next day. The next day, clip off the male flower and remove the petals so that you expose the pollen on the male anthers. Open each female flower in turn, and use the male flower like a paintbrush to distribute the male pollen onto the female stigma. Use more than one male plant flower to pollinate each female flower. When you have finished painting the pollen, retape the female flower. It will go on to form a fruit with fertile seeds.

When the squash or pumpkins are ripe, remove seeds by scooping out the insides, as you would for making a jack-o-lantern. Put all of the stringy material along with the seeds and water in a clean 5-gallon bucket and leave it overnight. The viable seeds will sink to the bottom and the unfertilized seeds will float to the top. Rinse the viable seeds thoroughly and then dry, bag, and label them.

Straw bales add to the mix of raised beds in this community garden, where fellow gardeners can help themselves to a little extra mulch if needed.

7 PERMACULTURE AND COMMUNITY

Fair Share

MY FAVORITE PART OF permaculture gardening is when we share our bounty with friends, family, and community. I regularly host work days at the Permaculture Institute of the East Bay (PIE), and it's a great way to share food, plants, and seeds, to learn new skills, and to create community around a shared task. Just as with your garden crops, these occasions are often filled with delightful surprises. One of the easiest and most rewarding ways to share with others is to organize a volunteer day for a large-scale project, either in your yard or in a community space.

Many hands truly do make for light work. Most of us don't have the time and space to grow everything, so select seed crops and edibles you are passionate about. Once you have harvested, processed, or stored yields from your garden, you can share them with others in your community. Sharing food and seed for future harvests is guided by the ethic of fair share and the permaculture principle of integration not separation. And you can also contribute to the effort to preserve heirloom varieties and maintain crop diversity.

SEED LIBRARIES

Along with some friends, I started the Bay Area Seed Interchange Library (BASIL) more than twelve years ago. BASIL is hosted at the Ecology Center in Berkeley, which has been a hub of ecological and social justice activism for more than forty years. Their mission statement reads, "BASIL is a free, community-based, urban seed project committed to disseminating and celebrating local varieties of seed stock and raising awareness about the importance and relationship between biological and cultural diversity." In addition to a bookstore, BASIL maintains a small corner with drawers and shelves full of home-grown seeds. This seed library is similar to a book library. Thousands of people come through and check out seeds for the season and then return them through an annual seed swap.

While BASIL is the veteran grassroots seed-saving group, other people have been inspired to start local and accessible seed-saving movements. Richmond Grows, co-founded by Rebecca Newburn and Catalin Kaser, is a free seed exchange hosted in the Richmond Public Library in Richmond, California. Library users get a small video tutorial on how to use the seed library and take only small amounts of easy-to-save seeds. To reflect the diverse community of Richmond, the whole process has been translated into Spanish and Mandarin. The library staff has let Richmond Grows start a small community garden adjacent to the building to put those seeds to use.

Rebecca's great organizational skills and enthusiasm have led the way for six other seed libraries founded this year in the Bay Area alone. Seed libraries are sprouting up in states across the country, including Colorado, Connecticut, Illinois, Kansas, Michigan, New Mexico, Oregon, and Pennsylvania.

You can also share seeds with friends and neighbors in a more informal way. A great benefit in having people come to your garden and learn to process seeds is that seeds are usually spilled at some point during the threshing and winnowing process, and next season a patch of arugula, calendula, or lettuce pops up in the paths and grows just fine without any help. The challenge is for you to recognize the seedlings that aren't weeds, and then see if you can live

with wherever they've decided to emerge. If you know they'll get trampled where they are, simply transplant them to a garden bed, or pot them and place them on the patio.

COMMUNITY FOOD FORESTS

There are a growing number of food forest gardens on a community scale too. One that caught my attention was at the Boys and Girls Club in Petaluma, California, where one hundred and fifty engaged citizens teamed up to create a community garden over the course of more than three days. They replaced the 3000-square-foot lawn, which needed 80,000 gallons of water annually to keep it green, with layers of cardboard, compost, and woodchips. By digging swales, redirecting the roof water, and installing drip irrigation, they reduced water usage by 80 percent.

These gardeners mimicked a forest ecosystem by planting in layers, with thirty fruit trees and eighty native habitat plants, medicinal plants, and nitrogen fixers. The leftover clay from digging the swales was made into a pizza oven and bench across the street, demonstrating the principle of producing no waste. Since the big blitz of establishing the

Richmond Grows' website, RichmondGrows.org, has open source software that can be used by anyone to start a seed library. It's easy to replicate the model, and more than a dozen other seed libraries have done so.

garden, the club has held many work parties, seasonal celebrations, and tours to engage the public. They also host workshops on topics such as pruning, grafting, and installing rainwater catchment systems.

COMMUNITY SUPPORTED AGRICULTURE

Community Supported Agriculture (CSA) is a model of supporting local farmers who grow food in a responsible manner. Members provide money to farmers for the growing season, and in return they get a weekly share in the harvest, either delivered to a central spot or, sometimes, directly to the doorstep. In addition to fresh, seasonal produce, many CSAs also offer meat, poultry, cheese, honey, and other grocery items. Local Harvest (localharvest.org) lists many CSAs across the United States and Canada, as well as farmer's markets, family farms, and other sources of locally grown food.

VOLUNTEER DAYS

Barn-raising used to be common in rural communities, with everyone coming together to share the work of building a barn. We now call this type of communal reciprocity "volunteer day," which is a great way to gather people to share the tasks of growing and harvesting food. And better yet, it's a terrific way to see all of the permaculture principles in action.

How do you organize a volunteer day? Doing physical work together like spreading woodchips or laying down compost makes people hungry, so you can organize food sharing through a potluck. Potlucks are a symbol of the abundance our gardens bring forth. I schedule volunteer days at my garden around activities like seed sharing and harvesting, and I always advertise them as potlucks, so everyone ends up sharing and

My community is teeming with organizations that are helping to ensure food security and social equity through growing and distributing healthy food. Phat Beets (phatbeetsproduce.org) supports local farmers and farmers of color, and has a partnership between a youth market garden and a local hospital obesity prevention program.

eating a greater variety of food. I have extra plant starts from my nursery to give away at these events, and there's usually a patch of kale, chard, or collards that can be harvested from the garden for the volunteers. I encourage people to graze around the garden while they work, and they always seem to end up near the blackberries, raspberries, and strawberries.

One or two energetic people can dig a lot of swales or move a truckload of woodchips, but it's more fun to have lots of people in on the project, learning from each other and getting a sense of accomplishment in only a few hours. Here are a few tips to keep things organized.

- Make it fun and educational.
- Have a clearly defined purpose and goal.
- Figure out the project details ahead of time.
- Have various projects for people with different skills and abilities.
- Provides tools and gloves.
- If the event takes place at a community garden or other site away from home, find the locations of nearby public bathrooms.
- Try to have something tangible that participants can take home, like plant starts or some crisp fall apples.
- Use social media like Facebook, Twitter, Yahoo, and Google groups to get the word out. Post before and after pictures on these pages after the project is completed.

Social media

Social media has found its way into nearly every area of our lives, and this includes gardening. There are websites, blogs, Facebook pages, and tweets on topics ranging from urban gardening to homesteading. There are also Yahoo and Google groups (and others, I'm sure) where interested gardeners can share information about specific things like how to sheet mulch or raise chickens, or to post events like volunteer days and garden tours. Permaculture encourages you to learn from traditional ways and adapt to local conditions, so take advantage of these updated methods of sharing information and creating community.

I'm part of an East Bay Yahoo group of nearly eight hundred members. Every week we post workshops, films, speakers, tours, work parties, volunteer days, and more. There are hundreds of groups like this across the country, so join one in your area or start your own. Don't look solely for permaculture groups—many other sustainable agriculture traditions can help you learn about growing food. Search online with keywords like biodynamic, biointensive, organic gardening, and urban homesteading.

PRODUCE SWAPS

Produce swaps, or crop swaps, are becoming more popular across the country and are a rewarding way to help build community. They're held in public places with easy access, so even those who don't have anything to offer one week may be inspired to bring something to share the next week. Everyone piles their garden abundance on tables: carrots, potatoes, chard, beets, purple string beans, vegetable starts, plums, lemons, honey, herb bundles, and a lot more. People begin by taking just a few things, and then when it is clear there will be plenty for everyone, can take more until it is all gone.

If there are no crop swaps in your area, consider starting one. Gather a few local gardeners who have extra produce to swap, find a central public location, post flyers, and use social media to spread the word; then let the swapping begin!

To start your own crop swap, it's best to start small but be consistent—always have the swap on the same day and time so it can build momentum.

Transition Berkeley
CROP SWAP
Mondays 6:30-7:30

TRADES AND BARTERING

Trades and bartering are a way to keep your money in your local economy. In permaculture, this is called alternative economics. The idea is to trade a skill you have with someone who is skilled in another area. Years ago, I traded my gardening work for 20 yards of compost. More recently I helped a friend plant the understory of his food forest; in exchange, he installed a woodstove at my house.

On a larger scale there are more formalized barter networks. The Ithaca Hour, based on the local economic trading system (LETSystem) in Ithaca, NY, gives everybody's work the same value. One hour of gardening, for instance, earns an hour that you can spend on services like a haircut or computer help. You can find out more about the LETSystem, along with their design manual, at their website, gmlets.u-net.com.

350 HOME AND GARDEN CHALLENGE

You can see the impact on one household when you transform your yard with permaculture, so imagine what can happen when we get together and help each other with our individual projects. In 2011, Sonoma County joined with the nonprofit organization Daily Acts to sponsor a 350 Home and Garden Challenge (the number 350 represents the number of parts per million of carbon needed to stabilize climate change). Citizens replaced hundreds of turfgrass lawns with food gardens, built twenty-one greywater systems to save water, and much more. Thousands of people got together to make this dream a reality—a good example of the power of positive local action. For more about the international 350 campaign founded by author Bill McKibben, go to 350.org.

Creating a Neighborhood Assessment

A neighborhood assessment is a great way to start organizing all the local resources available to help you create community around food sharing. The neighborhood assessment helps you identify nearby supportive people and resources so you don't have to reinvent the wheel. It's also a way to feel connected with, and more knowledgeable about, your community.

First, make a map of your neighborhood and mark the nearest farmers' market, CSA providers, food co-ops, grocery stores with local products, community gardens, and social justice gardening organizations. Next, list all the neighbors you know and make a goal to meet at least one new neighbor. Also contact any existing neighborhood or community organizations.

For emergency preparedness, identify where other people are growing food (veggies, fruit, bees, chickens, etc.) or harvesting greywater or rainwater. Is there a public safety/disaster preparedness group in your neighborhood? If not, consider starting one.

Include local schools on the map, and whether any have a school garden. Look for unused land and vacant lots that have potential gardening space. Is there any fruit or other food that is not being harvested? Include a short description of these places in your neighborhood assessment.

Teaching Abundance

Community gardens encourage us to eat local food, learn more about how our food is grown, and create a network of like-minded people. Community gardens are growing in popularity across North America and many have waiting lists. Check with master gardeners or your local municipality to find a list of community gardens. If you get involved, bring your newfound permaculture knowledge with you!

Three great examples of community farms and gardens are City Slicker Farms in Oakland, California (cityslickerfarms.org), the New York Permaculture Exchange (permaculture-exchange.org), and Growing Hope in Ypsilanti, Michigan (growinghope.net). By pairing local needs with local solutions, these organizations create stronger communities, help put healthy food on more plates, and empower participants with food and gardening knowledge.

Based in West Oakland, City Slicker Farms (CSF) is a nonprofit that runs seven community market farms where they teach gardening skills to youth interns and the wider community. The produce they grow is sold on a sliding scale, given away directly from the farms, or sold through local farmers' markets. City Slicker Farms shares a greenhouse with the Ralph J. Bunche Academy, a high school, where they raise fifteen thousand seedlings annually to provide food for their mini-farms, for a backyard garden program, and for sale to the public.

Each year, CSF organizes a mix of volunteer and paid construction workers to build twenty new small-scale backyard gardens a year. Polluted soils are a problem in industrial areas like West Oakland, so they install raised beds. An ongoing gardening mentoring program trains residents to become their own backyard farmers—a resilient strategy that helps the gardens succeed over time.

Another organization dedicated to improving the health and economy of West Oakland is People's Grocery. Not only do they distribute weekly produce boxes (Grub Boxes) to needy families, but they also manage a greenhouse enterprise program and garden at California Hotel, a low-income senior housing development.

For ten years, the New York Permaculture Exchange has been a hub of affordable and accessible permaculture education and action. Claudia Joseph is a permaculture teacher, consultant, community gardener, and founder of the New York Permaculture Exchange. She says, "Permaculture

The 350 Home and Garden Challenge, started in Sonoma County, California, by Daily Acts and partner groups, showed how much can be accomplished when community members, nonprofits, and governments join together to implement permaculture principles. The lawn outside the Petaluma Library was sheet-mulched and replaced with edible and non-edible permaculture plants. Find out more about community-scale permaculture solutions at dailyacts.org.

is more than just good gardening. It is a lifelong approach to analysis, actions, and social structures that results in better relationships, more functional systems and a clearer understanding of everything necessary for human communities to thrive."

In a demonstration of permaculture ethics, Claudia was instrumental in redesigning the collaboration between the Park Slope Food Coop in Brooklyn and the Garden of Union, a cooperative community garden operated since the 1970s. The co-op makes compost deliveries to the garden from their produce department, totaling about 14 tons annually. Four more gardens have been added to the system as the co-op has tripled in membership, creating a wealth of organic "waste."

It can be challenging to teach and practice permaculture in an urban center like New York City, where people are busy and land is expensive. Claudia tells her students, "Start where you are," encouraging them to find ways to use their training in their own neighborhoods. Students get permaculture gardening training in exchange for help with the heavy lifting, demonstrating the principle of integration rather than segregation.

In an effort to revitalize the Ypsilanti downtown and to make fresh food accessible to residents, Growing Hope helped start the Downtown Ypsilanti Farmer's Market. In 2006, the market grossed $22,000. By 2010, they had forty vendors and grossed more than $100,000. The market was one of the first in the country to let customers pay for their produce with SNAP (Supplemental Nutrition Assistance Program, formerly known as food stamps) through Electronic Benefit Transfer (EBT). These purchases now account for 20 percent of their annual sales. Thanks to a vibrant Michigan Farmers' Market Association, more than eighty markets now accept EBT/SNAP.

Growing Hope uses small and slow solutions. The organization teaches a lot of people to grow food, and empowers communities to make positive changes through economic justice and an emphasis on self-reliance. They help people in their community—particularly low-income families—start growing with easy-to-build and accessible 4-by-4-foot raised beds. They have been helped by some big supporters, like the former governor of Michigan, Jennifer Granholm, who signed the Cottage Food Law at the Growing Hope Center. This law gave the green light to backyard gardeners to grow food for value-added products like baked goods and jams. In an economically challenged state with high unemployment, this is a positive trend. Now, with support from Growing Hope, members of the community are raising vegetables, fruit, herbs, and making baked goods and crafts for sale.

All of these organizations demonstrate earth care by turning unused land into productive organic micro-farms, people care by providing tools for economic self help, and fair share by giving back to the larger community. These are the permaculture ethics that can inform our edible gardening efforts at all levels—as individuals, in our families, and in our communities.

Tomatoes and basil thrive in this unheated greenhouse at Growing Hope in Ypsilanti, Michigan.

Resources

Wildheart Gardens
(wildheartgardens.com)

Christopher Shein's garden design, building, and maintenance business. Find out more about my work, my biodiesel truck, rainwater and greywater, edible plants, native plants, garden designs, and contact information.

Animal, Vegetable, Miracle
(animalvegetablemiracle.com)

After writing a book detailing the year they spent trying to live as locavores, Barbara Kingsolver, Steven L. Hopp, and Camille Kingsolver started this website to continue their story. The site contains resources for finding local foods, recipes, and plenty of information on food security, sustainable agriculture, and policy issues.

The Apios Institute
(apiosinstitute.org)

Founded by David Jacke, the Apios Institute supports gardeners, farmers, and others who want to create ecosystems based on temperate climate forests. Contains lots of examples of polyculture plantings and food forests.

The Beehive Design Collective
(beehivecollective.org)

This activist design collective creates educational posters and artwork to raise ecological and social awareness.

The Cultural Conservancy
(nativeland.org)

This organization works to protect and restore indigenous cultures by bringing people back to their language, land, farming, and wild harvesting traditions.

Earth Democracy
(navdanya.org)

Author, activist, and scholar Vandana Shiva leads this Indian nonprofit center for biological and cultural biodiversity. Navdanya can mean "nine seeds" or "new gift."

The Ecology Center
(ecologycenter.org)

This organization started the modern curbside recycling industry in California more than forty years ago. They also run projects for sustainable living, including city farmers' markets, Farm Fresh Choice, a bookstore, a library, nonprofit structure shares, and Bay Area Seed Interchange Library (BASIL).

Gary Nabhan
(garynabhan.com)

Gary is an ethnologist and sustainable agriculture activist. His website is a wealth of information on place-based food and indigenous Sonora desert and Southwest slow foods stories and observations.

Greywater Action
(greywateraction.org)

This is a collaborative group of prime movers and shakers on the sustainable water front. The website has teaching tools on greywater recycling, rainwater harvesting, and human waste recycling systems.

Landscape Horticulture at Merritt College
(merrittlandhort.com)

Merritt College was among the first to offer a certificate in permaculture design. In addition to a one-acre student-run food forest, the college offers related classes such as mushroom cultivation, natural building, herbs, beneficial beasts, insects, and botanical drawing.

Movement Generation: Justice and Ecology Project
(movementgeneration.org)

A social justice activist organization, Movement Generation facilitates education and social justice, including liberation permaculture and strategic retreats for youth of color.

Native Seeds/SEARCH
(nativeseeds.org)

Based in the Southwest, this nonprofit seed company connects young farmers and the general public to heirloom seeds suited to low- and high-desert climates. Their mission is to preserve and disseminate ancient seeds and their wild relatives.

Oakland Institute
(oaklandinstitute.org)

Established by Anuradha Mittal, this think tank based in Oakland, California focuses on land rights, sustainable food systems, international aid, and more.

Occidental Arts and Ecology Center
(oaec.org)

Based in Sonoma County, California, this nonprofit center offers the longest continually running Permaculture Design Course in California. Teachers include Brock Dolman, and there are many collaborations, such as Doug Gosling's seed-saving gardens.

Permaculture Activist
(permacultureactivist.net)

The nexus of North American permaculture, this organization provides information on courses, books, events, and much more. Its quarterly publication includes regional news, special features, and a calendar.

Rainwater Harvesting for Drylands and Beyond
(harvestingrainwater.com)

Brad Lancaster is the guru of sustainable rain-fed gardens for dry climates. His three-part book series gives plenty of inspiration and technical information.

Starhawk's Tangled Web
(starhawk.org)

Starhawk is a global activist and rabble rouser—an eco-pagan witch, author, filmmaker, and co-facilitator of Earth Activist Training.

Urban Permaculture Guild
(urbanpermacultureguild.org)

This compendium provides permaculture information, from teacher training to design certificates, with plenty of interviews, videos, and great links, thanks to Kat Steele and her amazing networking skills.

Metric Conversions

FEET	METERS
1	0.3
2	0.6
3	0.9
4	1.2
5	1.5
6	1.8
7	2.1
8	2.4
9	2.7
10	3
20	6
30	9
40	12
50	15
100	30
200	60
300	90
400	120
500	150
1,000	300
2,000	610
3,000	910
4,000	1,200
5,000	1,500
6,000	1,800
7,000	2,100
8,000	2,400
9,000	2,700
10,000	3,000

VOLUME AND SURFACE

1 cup, measuring	8 fluid ounces
1 gallon (U.S.)	3.785 liters
	0.833 British gallon
1 acre	43,560 square feet
	0.405 hectare
1 cubic yard	0.765 cubic meter

TEMPERATURES

°C = $\frac{5}{9}$ × (°F−32)
°F = ($\frac{9}{5}$ × °C) + 32

INCHES	CM
¼	0.6
½	1.3
¾	1.9
1	2.5
2	5.1
3	7.6
4	10
5	13
6	15
7	18
8	20
9	23
10	25
20	51
30	76
40	100
50	130

Bibliography

Arora, David. 1986. *Mushrooms Demystified.* 3rd rev. ed. Berkeley, California: Ten Speed Press.

Ashworth, Suzanne. 2002. *Seed to Seed.* Decorah, Iowa: Seed Savers Exchange.

Bradley, Fern Marshall, ed. 1992. *Rodale's All-New Encyclopedia of Organic Gardening: The Indispensable Resource for Every Gardener.* Emmaus, Pennsylvania: Rodale Press.

Coleman, Eliot. 1989. *The New Organic Grower.* Chelsea, Vermont: Chelsea Green.

Conant, Jeff, and Pam Fadem. 2008. *A Community Guide to Environmental Health.* Berkeley, California: Hesperian Foundation.

Crawford, Martin. 2010. *Creating a Forest Garden: Working with Nature to Grow Edible Crops.* Devon, UK: Green Books.

Creasy, Rosalind. 2010. *Edible Landscaping.* 2nd rev. ed. San Francisco, California: Sierra Club Books.

Fukuoka, Masanobu. 1978. *One Straw Revolution.* Emmaus, Pennsylvania: Rodale Press.

Hemenway, Toby. 2009. *Gaia's Garden.* 2nd rev. ed. White River Junction, Vermont: Chelsea Green Publishing.

Holmgren, David. 2002. *Permaculture: Principles & Pathways Beyond Sustainability.* Hepburn, Australia: Holmgren Design Services.

Keith, Lierre. 2009. *The Vegetarian Myth.* Oakland, California: PM Press.

Jacke, Dave, and Eric Toensmeier. 2005. *Edible Forest Gardens.* White River Junction, Vermont: Chelsea Green Publishing.

Jeavons, John. 2006. *How to Grow More Vegetables.* 9th rev. ed. Berkeley, California: Ten Speed Press.

Jenkins, Joseph. 2005. *The Humanure Handbook.* 3rd rev. ed. Grove City, Pennsylvania: Joseph Jenkins, Inc.

Kaplan, Rachel, with K. Ruby Blume. 2011. *Urban Homesteading.* New York: Skyhorse Publishing.

Kourik, Robert. 1986. *Designing and Maintaining Your Edible Landscape Naturally.* Santa Rosa, California: Metamorphic Press.

Lancaster, Brad. 2008. *Rainwater Harvesting for Drylands and Beyond,* Volume 1. Tucson, Arizona: Rainsource Press.

Ludwig, Art. 2007. *Create an Oasis with Greywater.* 5th rev. ed. Santa Barbara, California: Oasis Design.

Lyle, Susanna. 2006. *Fruit and Nuts.* Portland, Oregon: Timber Press.

Margolin, Malcolm. 1975. *The Earth Manual.* Berkeley, California: Hayday Books.

Mars, Ross, and Jenny Mars. 2008. *Getting Started in Permaculture.* 3rd rev. ed. White River Junction, Vermont: Chelsea Green Publishing.

Martin, Deborah L., and Grace Gershuny. 1992. *The Rodale Book of Composting: Easy Methods for Every Gardener.* Emmaus, Pennsylvania: Rodale Books.

McClure, Susan, and Sally Roth. 1994. *Companion Planting*. Emmaus, Pennsylvania: Rodale Books.

Mollison, Bill. 1988. *Permaculture: A Designer's Manual*. Edited by Reny Mia Slay. Tyalgum, Australia: Tagari Publications.

Mollison, Bill. 1999. *Introduction to Permaculture*. 2nd rev. ed. Tyalgum, Australia: Tagari Publications.

Orsi, Janelle, and Emily Doskow. 2009. *The Sharing Solution*. Berkeley, California: Nolo Press.

Permatil. 2006. *A Resource Book for Permaculture: Solutions for Sustainable Lifestyles*. Bali, Indonesia: DEP Foundation.

Pinkerton, Tamzin, and Rob Hopkins. 2009. *Local Food: How to Make It Happen in Your Community*. Totnes, UK: Transition Books.

Stout, Ruth. 1963. *Gardening Without Work*. New York: Devin-Adair.

Toensmeier, Eric. 2007. *Perennial Vegetables*. White River Junction, Vermont: Chelsea Green Publishing.

Watkins, David. 1993. *Urban Permaculture*. Clanfield, Hampshire, UK: Permanent Publications.

Yamaguchi, Mas, and Vincent E. Rubatzky. 1999. *World Vegetables*. 2nd ed. Florence, Kentucky: Chapman and Hall.

Index

D

E

F

G

H

I

J

K

Q

R

S

T